REDISCOVER THE JOYS AND BEAUTY OF NATURE WITH TOM BROWN, JR.

THE TRACKER
Tom Brown's classic true story—th~~ ~~
magical high-spiritual adventu~~ ~~
Don Juan

THE SEARCH
The continuing story of *The Tra* ~~ ~~ ancient
art of the new survival

THE VISION
Tom Brown's profound, personal journey into an ancient
mystical experience, the Vision Quest

THE QUEST
The acclaimed outdoorsman shows how we can save our
planet

THE JOURNEY
A message of hope and harmony for our earth and our
spirits—Tom Brown's vision for healing our world

GRANDFATHER
The incredible true story of a remarkable Native American and
his lifelong search for peace and truth in nature

AWAKENING SPIRITS
For the first time, Tom Brown shares the unique mediation
exercises used by students of his personal Tracker Classes

THE WAY OF THE SCOUT
Tom Brown's real-life apprenticeship in the ways of the
scouts—ancient teachings as timeless as nature itself

THE SCIENCE AND THE ART OF TRACKING
Tom Brown shares the wisdom of generations of animal
trackers—revelations that awaken us to our own place in
nature and in the world

AND THE BESTSELLING SERIES
OF TOM BROWN'S FIELD GUIDES

Berkley Books by Tom Brown, Jr.

THE TRACKER (as told to William Jon Watkins)
THE SEARCH (with William Owen)
TOM BROWN'S FIELD GUIDE TO
    WILDERNESS SURVIVAL
    (with Brandt Morgan)
TOM BROWN'S FIELD GUIDE TO NATURE
    OBSERVATION AND TRACKING
    (with Brandt Morgan)
TOM BROWN'S FIELD GUIDE TO CITY AND
    SUBURBAN SURVIVAL (with Brandt Morgan)
TOM BROWN'S FIELD GUIDE TO LIVING
    WITH THE EARTH (with Brandt Morgan)
TOM BROWN'S GUIDE TO WILD EDIBLE AND
    MEDICINAL PLANTS
TOM BROWN'S FIELD GUIDE TO THE
    FORGOTTEN WILDERNESS
TOM BROWN'S FIELD GUIDE TO NATURE
    AND SURVIVAL FOR CHILDREN
    (with Judy Brown)
THE VISION
THE QUEST
THE JOURNEY
GRANDFATHER
AWAKENING SPIRITS
THE WAY OF THE SCOUT
THE SCIENCE AND ART OF TRACKING

About the Author

At the age of eight, Tom Brown, Jr., began to learn tracking and hunting from Stalking Wolf, a displaced Apache Indian. Today Brown is an experienced woodsman whose extraordinary skill has saved many lives, including his own. He manages and teaches one of the largest wilderness and survival schools in the U.S. and has instructed many law enforcement agencies and rescue teams.

Most Berkley Books are available at special quantity discounts for bulk purchases for sales promotions, premiums, fund-raising, or educational use. Special books, or book excerpts, can also be created to fit specific needs.

For details, write to Special Markets, The Berkley Publishing Group, 375 Hudson Street, New York, New York 10014.

TOM BROWN'S FIELD GUIDE TO LIVING WITH THE EARTH

TOM BROWN'S FIELD GUIDE TO LIVING WITH THE EARTH

Tom Brown, Jr., with Brandt Morgan

Illustrated by Heather Bolyn
and Trip Becker

BERKLEY BOOKS, NEW YORK

KETCHIKAN PUBLIC LIBRARY
KETCHIKAN, ALASKA 99901

The Publisher and Author disclaim any liability for injury that may result from
following the techniques and instructions described in the Field Guide, which could
be dangerous in certain situations. In addition, some of the techniques and
instructions may be inappropriate for persons suffering from certain physical
conditions or handicaps.

TOM BROWN'S FIELD GUIDE TO
LIVING WITH THE EARTH

A Berkley Book / published by arrangement with
the author

PRINTING HISTORY
Berkley trade paperback edition / November 1984

All rights reserved.
Copyright © 1984 by Tom Brown, Jr.
This book may not be reproduced in whole or in part,
by mimeograph or any other means, without permission.
For information address: The Berkley Publishing Group,
a division of Penguin Putnam Inc.,
375 Hudson Street, New York, New York 10014.

The Penguin Putnam Inc. World Wide Web site address is
http://www.penguinputnam.com

ISBN: 0-425-09147-3

BERKLEY®
Berkley Books are published by The Berkley Publishing Group,
a division of Penguin Putnam Inc.,
375 Hudson Street, New York, New York 10014.
BERKLEY and the "B" design are trademarks belonging to
Penguin Putnam Inc.

PRINTED IN THE UNITED STATES OF AMERICA

20   19   18   17   16

TOM BROWN'S DEDICATION:

To Mom Schill, Robin, Aunt Joy, and Gee Gee—my favorite in-laws; to Paul, my son, who taught me to be a father; and to Kelly, who taught me to see love in everyone.

BRANDT MORGAN'S DEDICATION:

To my father, Arthur Morgan, Jr., who loves the soil and knows what it is to be a caretaker.

SPECIAL THANKS TO:

The people of the earth who carry with them the ancient fires, burning deep within their souls. The people living in balanced harmony with Earth Mother and the greater vision beyond the self.

And especially to Judy and Tommy. They are the air that uplifts my wings and allows my spirit to soar like the eagles.

Warm thanks also to Eric Heline, Joe McDonald, Mike and Robin Clinchy, Frank and Karen Sherwood, Kurt Folsom, Michelle Kaestner, Mary Ramsey, Ruth Morgan, Ralph Panaro, Cynthia Lewis, Charlie Johnson, Jon Clark, Dave Boyd, and Lynn and Steve McDowell.

CONTENTS

TOM BROWN'S
FIELD GUIDE TO
LIVING WITH
THE EARTH

INTRODUCTION

There are many ways of looking at survival, but most people regard the concept only in a narrow sense. They imagine some poor lost soul stuck out in the woods, debilitated, and unable to fend for himself. Survival to most people means a bitter struggle against the "hostile" forces of nature, with the main goal being to get back to the comforting arms of civilization as soon as possible.

I don't see survival that way, and I never have. From the time I was seven years old, Stalking Wolf, my Apache Indian teacher, taught me to love nature as an extension of myself. He taught me that I would never be lost unless I had a place to be and a time to be there. He told me that I would never have to fear nature and that it would always provide for me as long as I lived in harmony with it and did not try to fight it. With this frame of mind, survival took on an entirely different meaning. Over the years the wilderness became my home and my sanctuary, and living there became as fruitful and rewarding a life-style as I have ever known.

I don't mean to say that survival doesn't require skill. During the ten years I spent with Stalking Wolf, I repeatedly discovered how important it was to be able to provide shelter, water, fire, and food before I could truly enjoy myself and feel at home in the wilderness. One of the main reasons I decided to write a series of survival handbooks was to help other people experience the same kind of joy and security I have felt in wild places. The first book in this series, *Tom Brown's Field Guide to Wilderness Survival*, explains basic survival skills. The second book, *Tom Brown's Field Guide to Nature Observation and Tracking*, expands on some of these skills and points the way toward a deeper connection with nature.

In a way, this book is a logical extension of the first two. It deals with advanced survival and nature awareness skills that have been used by our ancestors for thousands of years. Most of these are aimed at providing necessities for long-term living rather than mere subsistence living. They represent a step up from basic survival to a greater level of comfort and convenience. Included are construction of semipermanent shelters and simple furniture, how to "carry" fire, how to make pots, baskets, mats, and fine bows and arrows. There is also more on the art of stoneworking and more on how to blend and flow with the woods.

These are all practical skills that can be used by people either in survival situations or on regular camp-outs or wilderness treks. Like the basic survival skills, they can add to a sense of security in the woods. As advanced skills, they can also instill a greater pride in craftsmanship. But most importantly, they can teach some precious lessons about the art of living with the earth.

Living with the earth in its fullest sense is much more than mere survival. It is also more than the frenzy and desperation that passes for life in much of the modern world. Living with the earth is an intimate belonging, like the connectedness of a well-rooted tree. There is a solidness to it. There is also growth. And that growth is bound up with soil, water, and sun in such a way that there is no separation. The tree is part of the earth and the sky. It protects and nourishes other life. It lives in harmony with everything around it.

It is that sense of harmony and belonging that interests me most. Like trees, we also need solid roots. We need to feel connected, not only to people and places, but to other animals and plants and even to entities that most of us consider lifeless. Stalking Wolf and the wilderness have taught me that water, sky, and earth support people more directly than their own legs or their monthly paychecks. Realizing this connection—and better yet, living it—can be one of the most profound and rewarding experiences life has to offer.

The skills presented in this book, then, are not only meant to be practical survival skills, but windows to a deeper awareness. They are all primal skills; that is, they have been used by all people who live close to the earth almost since the dawn of humanity. To such earth people these crafts are not only ways of life, but expressions of their innermost beings. Each embodies the skill of the crafts-person, as well as a great deal of love and devotion. Each one in its own way is also an expression of the spiritual roots that bind all earth people to the soil and the sky and to each other.

Most of our lives are far more complicated than those of our ancient ancestors. But beneath our civilized facade we are strikingly the same. We have the same needs, the same instincts, and the same drives. The main difference is that our senses are duller. But we can retrieve much of the sensitivity that has been lost.

So often we learn only the superficial things about our ancestors. We read about their cultures as though they had only the most distant relation to our own. We look at museum exhibits about primal peoples and think of their artifacts as relics of worn-out societies. It rarely occurs to us to look for deeper meaning and wisdom in these "relics." There is much more to them than meets the eye. And much of it can only be discovered by doing—ideally, by reconstructing both the artifact and some of the conditions under which it was made. If we do this, we can become part of a society vastly larger than the one we live in. It is a society not only of people, but of plants, animals, rocks, clouds, and spirits—a rich and fascinating community of time that stretches back to our very beginnings.

I do not mean that we should all go back to living in caves. Nor do I mean that survival is the only way to get back to nature. Brief and occasional trips to the woods can have a powerful effect on the psyche. Even a walk in the park can help to awaken old instincts and intuitions. But any person who lives for a month on an acre of wild land will never be able to turn that land into a parking lot. That person will know his need for air and water and earth as never before. He will be more aware of how his decisions affect his surroundings. He will know why a chunk of wilderness he has never seen is as much a part of him as the property he has bought and paid for. He will remember his connection and live more deliberately, whether in a wilderness cabin or a high-rise apartment building.

Not everyone who reads this book will be interested in living off the land for a week. Not every reader will want to make a bow and arrow or an earthen pot or a fiber rug. But such skills have important lessons for us all. Wild nature is in everyone. We are all tied to the soil. No matter how fragile or distant the connection may seem, it is still there. Even a fleeting acknowledgment of that fact can make a difference in the quality of our lives.

About the Skills

The skills in this book are adapted from a variety of traditions. Many of them are based on native American methods, partly because I have been so heavily influenced by them myself and partly because most people on this continent are more familiar with them. But I want to emphasize that the native Americans are not the only ones who used these skills. They have been used by native peoples all over the world, in all ages and places, in response to the changing environment.

Each culture's methods differ according to its own needs and materials. The tipi and the igloo have very similar functions, but their design and construction are vastly different because of the stark differences in the environment. The bows that were used by the Plains and Eastern Woodlands Indians had similar functions, but their designs varied greatly because of the differences in game and hunting styles among those tribes.

By the same token, the skills presented in this book are very similar to those of many earth peoples, but they do not follow any one tradition. The reason for this is simple. These tools are not intended to be museum pieces or dead reminders of bygone cultures. They are intended to be living, working tools that can be used by modern people in a wide variety of places and circumstances. There is no point in making a Plains Indian bow in the Sioux tradition unless you have access to a horse and a herd of buffalo. In storm country there's no point in making a desert shelter that doesn't keep out the wind and rain. You have to adapt it to fit the circumstance.

Adaptation comes through experience. I have discovered, for example, that I don't like the long Seminole fish spear. Nor do I like the shorter Penobscot spear. The one that works the best for me is a medium-length spear that is a mixture of three different traditions. Similarly, the bows described in this book are designed from a mix of different traditions, plus my own experience. The shelters described here are also a mix of traditions, but I have found through experience that they can be adapted to almost any part of the country.

Neither am I interested in cataloguing the many different ways of making things. Other books have already done that. There are thousands of different designs for pots, bows, bowls, baskets, mats, furniture, and other advanced survival tools. What I want to pass on is something that *works*—something that has a good success ratio in a survival situation and that can be made with a minimum of difficulty. For this reason I have, in most cases, explained only one or two methods and designs for each skill. Once you understand these methods, you can read other books, go to museums, and combine traditions and ideas to modify them in any way you like.

When it comes to survival, adaptation is the name of the game. Even the native Americans were not purists. They, too, took ideas from other cultures and used their minds to create better extensions of their limbs and senses. After the arrival of the Europeans, some of the native Americans found uses for European tools. Many Indians even abandoned their stone quarries and began making knives and arrowheads from the bottoms of cast-iron frying pans.

Civilization has its problems, but it also has its advantages. My aim is not to get you to give up modern conveniences and ways of thinking, but to experience the beauty of the ancient tradition of living with the earth. The point of survival living—or, as this book suggests, "earth living"—is not to abandon the modern world, but to live more fully within it. I am convinced that part of that fullness comes with simplicity and at least a marginal connection to our ancestral roots. It doesn't take much to reestablish the connection. I hope that this book suggests some rewarding ways of doing it.

1
EARTHMIND

It may be that some little root of the
sacred tree still lives. Nourish it then, that it may
leaf and bloom and fill with singing birds.

Black Elk

The sacred tree of which Black Elk spoke long ago is the tree of awareness that has been given to all of us. Far removed from the soil as most of us are, we are all people of the earth. Our roots reach back beyond the asphalt and concrete, down deeper than any building foundation. They reach back to the very dawn of creation.

Hidden in our hearts are levels of awareness we have forgotten. Like great reservoirs they lie just beneath the surface, waiting to be tapped, waiting for their chance to gush to the surface and remind us of who we really are. We can all see, hear, smell, taste, touch, and feel much more than we do. The roots of the sacred tree are as much our birthright as the American Indian's or any other people who have walked the earth. They only wait to be rediscovered.

Seeing Beyond the Surface

The road to reawakening begins with seeing our heritage for what it is. Nobody has taught me more about this than Stalking Wolf, and he did it just by living. He didn't have to preach or explain things. He didn't have to push and prod. His method was a quiet one, and, like the way he lived, it sank deeper into my consciousness than all the books I have ever read.

When Stalking Wolf looked at a tree, he did not just see a trunk with branches and leaves; he saw *into* the tree. He noticed the texture of the bark, the twist and flow of the limbs, the way the tree swayed in the wind, how its leaves fluttered, what birds and animals had been using it, whether its fruit was ripe—even what kinds of insects crawled on it. He saw all these things because the tree had survival value. He knew that the tree's fruit provided food; that its leaves and branches offered shade and shelter; that its bark would give fiber for mats, baskets, and ropes; and that from its wood he might make a variety of tools and implements. To him, the tree was a pulsating fountain of life.

Stalking Wolf's observations led to important discoveries. Bird voices told him where to find the nearest fox or owl. The wind told him how to approach a deer or rabbit. The acorns told him whether the hunting would be good next year. Everything in the environment was rich with survival information. But Stalking Wolf did not observe anything just by itself, because for him nothing existed in and of itself. While he looked at a tree, his senses were also reaching out to everything around it—to the forest, the grasses, the sky, the mountains, and even to the far horizon and the stars. Everything was related. Everything was part of the great cosmic unity, the spirit-that-moves-through-all-things.

Stalking Wolf also had a great desire to get to the core of things. This desire went far deeper than survival. When he looked at a tree, he could see its beauty and utility, but he also wanted to touch its spirit. He did not try to define the tree. He did not identify it and pretend that the tree's name told him everything there was to know about it. He realized he could never know it completely, and that it would always be somewhat mysterious. He also saw it as changing and growing. He knew the tree he saw today was not exactly like the one he saw yesterday, and that tomorrow it would not be the same as it was today. He saw everything in a process of becoming something else, and he gave everything room to grow and change in its own way.

This way of looking at things is not so unusual if we stop to think about it. Plants and animals have needs very similar to our own. We have all sprung mysteriously from the same earth, and we all have a common umbilical cord. The earth is our mother—the womb of creation and the giver of life. When we begin to sense the truth of this, those things that grow from the earth can then be seen as brothers and sisters. They are no longer "dumb creation," they are part of a complex and interdependent family. To wantonly take a life or waste or destroy anything is unthinkable, because it means wasting or destroying part of our family and ourselves.

"I am sorry, my brother, but I must take your life so that I may go on living." That is a very profound statement, but on some level that is what Stalking Wolf said when he took other life.

Why make such a statement? Because it puts death in perspective. It acknowledges the sacrifice and the gift. It serves as a prayer that connects us to the spirit of our "brother" that gave up its life for us. It casts a sense of reverence and appreciation over any act of taking, and it reminds us to use the gift in a responsible way.

From this point of view, ecology takes on a deeper meaning. The deeper meaning does not come from biology books or from dissecting and analyzing things. It comes through hunger and thirst. It comes through watching, smelling, listening, and feeling with all the senses, and through living as one of all the interlocking parts. It comes through interacting directly with sun, water, clouds, rocks, plants, and animals—through lying close to the ground and feeling the heartthrob of the earth.

Survival is not everything. Nature is also a great teacher. When I was very young, one day I wanted to know how Stalking Wolf could tell there was an owl in a tree without even looking up, and he said, "Go ask the mice." There was so much teaching in that simple statement, and even more in the experience of trying to unravel its meaning. I spent days and weeks down on my belly, watching mice. In the process, I finally learned that Stalking Wolf had been able to tell there was an owl in the tree by what the nearby mice did. But that was not all I learned. By spending time with the mice and opening myself up to what they had to teach, I not only learned about owls, but about coyotes, foxes, plants, grasses, spiders, snakes, and everything else in their world that I had the patience to observe and open up to.

At first glance, this kind of observation might not seem much different from that of a scientist, but there is a big difference. Stalking Wolf did not treat things as separate objects; he treated them as extensions of himself. He wasn't so interested in how things work; he wanted to know what was at the heart of them. He looked at things from the inside out—as part of the landscape and part of the universal life force. There is great awe and reverence in this kind of observation. In fact, it is a kind of prayer.

We all have a natural reverence for life. Deep down in the roots of our sacred tree, we know we cannot live without taking other life. We also know we cannot treat other animals as inferiors, because so many of them are gifted with senses and abilities ar greater than our own. We do not have the swiftness of the swallow, the ears of the fox, or the eyes of the eagle. But we can humble ourselves and learn from our animal teachers. If we watch the heron it will teach us how to be patient. If we watch the weasel it will teach us how to stalk and hunt. If we watch the chickadee it will teach us how to confront the storms of existence with good cheer. Every entity has its lessons, no matter how small or insignificant it may seem.

When I was young, I watched Stalking Wolf constantly to learn anything he might have to teach, and I noticed that he often treated the earth as though it were a living being. I asked him if this were so, and he said, "Grandson, can you not see the way of things? Does your life not come from the earth? Is the flesh of the deer not the flesh of your flesh? Is not your own blood the blood of Earth Mother? Is not your breath also the breath of the wind?"

It was not the words so much as how he said it. He said it with total conviction—as though it were something beyond doubting. I could not argue with him, and even then I knew in my heart that he was right. The years have only deepened that conviction.

Conversing With Life

I am no longer surprised that Stalking Wolf was always talking to the earth. To him, everything was alive and pulsating, and he wanted to connect and communicate with it. Sometimes this was a conversation of words, but more often it was a conversation of the heart. Once I watched Stalking Wolf go down to the creek to take a bath. He sat near the edge at first, quietly watching the water striders and reflections on the surface. At times his awareness seemed to drift with the current, traveling up and down the creek, then sinking to the bottom where little pickerel waved among the grasses.

After a while, he reached his cupped hands into the water and lifted them, dripping, to his face. In ecstasy he watched the droplets fall from his fingers and splash onto the surface, sending concentric ripples to the edges of the creek. He drank, and without words he talked to the water. He savored it. He reached out to it, feeling its surface tension with his hands and absorbing its hidden qualities with his mind and heart.

Finally Stalking Wolf stepped into a clear pool and slowly descended into the water until his body was almost submerged. He moved almost as one with the creek, and he seemed radiant with joy. He washed his face and body, feeling the cool smoothness of the water as it soaked into his skin. Finally he let the water cover him completely, and when he came to the surface again he relaxed and opened his eyes. His arms and legs dangled in the current, and as he looked up at the trees and open sky, he seemed to be suspended in the very womb of creation.

Stalking Wolf did not always pray openly, but sometimes he could not contain himself and often I could not tell the difference between his praying and his living. He was always expressing his love and appreciation for the entities of the earth, and it always came back to him.

Natural joy and awareness are not reserved just for people like Stalking Wolf. We can also immerse ourselves in the pools of existence. We can also see beyond the surface of things. We can do it by reaching out with our feelings—by realizing that all things have a common umbilical cord, and by opening ourselves up to receive answers through it. Sometimes these answers will come in the form of powerful feelings. Sometimes they will come in the form of dreams or visions. Other times they will come in flashes of intuition or insight.

There is little difference between the intuitions of primal peoples and ourselves. The dreams and visions that were given to the Wise Ones, the medicine people and prophets of old, are also available to us. But so often when something "supernatural" happens to us or we have a hunch or a flash of insight, we tend to discount it, saying, "Oh, it couldn't be," or "It probably doesn't mean anything," or "It was only a dream." Stalking Wolf did not limit his perceptions. He did not doubt his dreams and visions. For him, the world just beyond sight and sound was just as real as any other.

Most of us have been taught all our lives to believe only those things that can be argued logically or proved scientifically. We are taught that if we can't see something or analyze it in a test tube, it doesn't exist. We're always trying to explain and define things. We screen out and limit the subconscious, and in doing this we cut ourselves off from the mystery.

If we let things happen more naturally, we could be guided by a beautiful blend of logic and intuition. We do not have to explain and define everything; we can be more content just to experience and live with things. We do not have to break things up into little pieces; we can see how they fit into the bigger picture. There is little separation between physical, mental, and spiritual events. When all the levels of consciousness are wide open, everything is bound up together. Dreams are not just dreams, but profound experiences. All things are ultimately cloaked in mystery. Physical reality, dream reality, and spirit reality are all parts of the same great puzzle.

Even death loses its finality when seen in this way. Stalking Wolf knew that however far removed from his native land he might be, he was never separate from the rest of his tribe and the rest of humanity. To him, life was communal and cyclical. The playfulness of childhood, the strength of adulthood, and the wisdom of old age all combined to give direction and meaning to his life. He was never alone. Every separate thing had an important place in his life. The rocks and dirt gave him a firm foundation and a place for his roots to grow. The plants and animals gave him a sense of brotherhood and belonging. The water was his lifeblood. The air was his breath. The Great Spirit flowed through him as naturally as sap through the broad-spreading branches of a tree.

It has taken me many years to understand the true nature of this "tree," and how far its branches actually spread. I was often amazed, for instance, when Stalking Wolf could locate animals without seeing or hearing any sign of them. I once asked him about this and he answered: "Can you not feel an ant crawling on your flesh? Is not the flesh of Earth Mother the same flesh as yours? Are you not part of the animals and plants and all other things that are part of Earth Mother?"

I could not answer him then, but I know now that this is the real meaning behind the sacred tree. This is the real reason Stalking Wolf felt so tied to plants and animals and even so-called "dead" things such as rocks and clouds and water. It is not just because he was physically dependent on them, but also because he felt the universal umbilical cord that fed his spirit through them, and he was continually celebrating the great mystery of it all.

The Sacred Hunt

This celebration was not just something that Stalking Wolf did in his leisure time. It extended to everything he did, and in no part of his life was it more important than in the act of hunting. I am not talking about a hunt just for an animal or a plant, but for anything—a deer, a bowstave, an arrow shaft, even a mouthful of water. It was an attitude that permeated his whole being.

For Stalking Wolf, the sacred hunt began in the heart. It began with the realization that he had come from the earth and was part of the earth and everything on it. It began with the knowledge that he moved within the realm of all creation and that all of creation moved within him. There was no inner or outer dimension. With this knowledge, he could never think of owning or improving on the land; he only wished to live in harmony with it. He knew it was impossible to live without taking other life, so he aimed to live in a responsible way. He knew that whatever he did to the earth would have an effect on everything else, including himself.

For this reason, Stalking Wolf lived very deliberately. When he went out to take something from the landscape, the first thing he did was clear his mind. He would empty himself of all thoughts and slow himself down enough to feel his own rhythm alongside the rhythm of the spirit-that-moves-through-all-things. Then he would ask himself,

"Do I really need this deer or this bowstave? Is it important to my survival or well-being?" Then he would ask the Great Spirit whether it was all right to hunt for that entity.

If he felt that it was right, Stalking Wolf would begin the preparation. This might be simple or elaborate, depending on what he was looking for. If it was a plant, he would silently pray to the spirit of that plant, asking its permission and explaining the need for it and what it would be used for. If it was a deer, he might fast for several days. He would do this not just to hone his senses and camouflage his scent, but to know hunger and to show the spirits that his need was real. His devotion put the hunt into perspective.

The physical act of hunting was also very deliberate. It combined reverence and devotion with an intricate awareness of the landscape. When Stalking Wolf went out to find a bowstave, he never looked for the healthiest tree; he looked for a tree that had been crowded and kept from the light. He wanted one that had had to fight for its life, so that the growth rings in the trunk would be packed closely together. This would give more strength and snap to the bow. By taking a tree that was bound to die and that was only crowding its healthy brothers, he was helping himself and Earth Mother at the same time.

As he did this, he always prayed: "I am sorry, my brother, but I must take your life so that I may live." But the prayer was not just an apology. It was also a thanksgiving and a way of reassuring the tree's spirit that it would be put to good use. It was a way of acknowledging the gift and showing his respect and love for all of Creation.

Stalking Wolf's promises were not given lightly. Even after the hunt was over, he still felt an obligation to make good use of whatever he had taken. Nothing was wasted, and whatever was made from the offerings of the earth was made with loving care. This was not only for the sake of survival (since a shoddy job sould show up in the implement and eventually hurt his own livelihood), but to give thanks for the gift. When Stalking Wolf took a deer, he was very careful not to anger the Creator or the spirit of the deer because he firmly believed that the animal's spirit would tell the other animals and that no deer would want to offer itself to him again. When he made a bow, he was careful to stay in the right frame of mind, without any negativity or ill will, so that the "medicine" he put into the bow would help it to shoot straight and true.

This reverence was not reserved just for physical things; it extended to all things. Stalking Wolf did not start a fire without saying a silent prayer, and he did not sit by a bed of coals without talking to the spirit of the fire. Even when he wasn't talking, I could see the reverence reflected in his eyes. Like most earth peoples, he was in constant conversation with the Great Spirit.

I learned most of Stalking Wolf's lessons very slowly. But every once in a while many lessons suddenly fell into perspective for me. One of these times was when I killed my first deer. I had tracked it for a week. I had lived with it, moved with it, and slept near it. I had looked into its eyes and talked to it with my heart. I knew almost everything about it. When the time finally came, I dropped from a tree and killed it with my knife and my bare hands. As I took the life from it, watching it ebb and flow between my fingers, I could see the terror in its eyes. But beyond the terror I could see the awesome gift that any animal gives when it gives up its life for another.

When it was over, I was covered with blood and very shaken and hurt. I felt I had just killed a member of my own family. I walked back to camp with the deer slung over my shoulders, a very broken young man. As I walked up to Stalking Wolf, still sobbing, he smiled at me and said, "Grandson, when you can feel the same way about a blade of grass that is plucked from the earth as you do about your brother deer, you will truly be one with all things." That is the meaning of the sacred hunt.

Living in the Moment

How can we nourish the sacred tree? Along with a reverence for life, few things can help more than living in the moment. I was reminded of this recently while looking at a beautiful cedar sculpture that sits in a museum at the University of British Columbia in Vancouver. In smooth, flowing curves the sculpture tells the legend of creation of the tribes of the Northwest Coast. It shows a huge raven with protective, outspread wings standing on a large clamshell. The raven is calling to the fearful humans inside the shell, encouraging them to come out and play. In this myth, the clamshell is like the earth itself—the womb of creation that protects and cradles all its offspring. Peeking from the inside of the half-open shell are humans of all ages, arranged in a circular fashion that suggests the cycle of life.

To me, this sculpture says the same things Stalking Wolf always said: Life does not travel in a straight line from birth to death. Creation is not just something that happened in one place and time. Life travels the sacred circle, and there is no ending. Creation is happening everywhere and in all time, endlessly emerging and moving in the midst of the eternal Now.

In many ways it was easier for Stalking Wolf to live in the moment. His life was not dictated by the clock, but by the natural cycles of the seasons. His days progressed slowly from sunrise to sunset, season to season. Within each day he did the things he needed to do to stay in tune with himself and the earth. That is all he wanted or needed.

Most of us do not live this way. But the schedules and preoccupations of modern life don't have to keep us shut off from our ancestral roots. There are ways of getting back. There are ways of nourishing the sacred tree, even in the midst of modern society. It is mostly a matter of living deliberately. It is a matter of slowing down and taking the time to remember what is important. It is taking the time to consult with the inner voice, whatever we conceive it to be, and by breaking away from the shackles of time.

There are many ways to bring outselves into the eternal Now. For one person it might be meditation. For another it might be time spent alone in nature, or in a cathedral. For a third it might be a hobby or a sport. I don't claim to have the only answer. Yet I can say that for me, at least, reconnecting myself with the earth through age-old survival skills has been one of the most direct pathways to timelessness that I have ever found. It has been a way of remembering, in the rush of life, that the roots of the sacred tree still live. It has been a way of inviting some of the forgotten branches to "leaf and bloom and fill with singing birds."

2
EARTHSHELTER

The tribe always camped in a circle,
and in the middle of the circle was
a place called Hocoka, the center.
Tatanka Ptecila (Short Bull)

Once survival is assured, a shelter becomes more than just a place to stay warm and dry; it becomes a home. A good shelter provides not only comfort and security, but a sense of permanence and a firm foundation for life. Whether the shelter is a shack or a luxurious home, it is the place we go out from and return to each day. It is the place where we eat, sleep, and nurture our relationships with family and friends. It is the place where we can sink our roots into the earth.

When Stalking Wolf was in the Pine Barrens, he lived most of the time in a dome-shaped shelter that was similar to a wigwam. It was set among the scrub oak and pine, far off to the side of the main trail. There he often sat and smoked and meditated. Sometimes my friend Rick, Stalking Wolf's grandson, and I would visit him. As often as not, we found him sitting on the bare ground or on a simple reed mat.

Stalking Wolf liked sitting on the ground, and he encouraged us to do the same. He said that the earth gave off unseen powers, and that we could feel them if we stayed close to it and did not wear too many clothes. So this is what we did. We went without shoes and shirts whenever we could, and we enjoyed the bare earth.

We enjoyed other things about Stalking Wolf's shelter. In the middle of it he had a small hearth with a fire pit. Around the hearth were hung a few wooden cooking utensils and a few earthen pots he had made with clay from the river. He had also made a backrest, a stool, a workbench, and a few other simple pieces of furniture. Herbs, roots, and tools were always hanging from rawhide cords or beams. His bow and arrows were hung above his bed, and his few other possessions were stored neatly around the periphery of the circle under the little platforms that ringed the edge of the shelter.

One of the first things that struck me about Stalking Wolf's shelter (and almost every other native American dwelling I have seen since) was its roundness. Sitting inside it had a calming effect. No matter where Rick and I sat, we were part of a circle that radiated from the central fire. This seemed natural to us, because we could see that most things in the universe tried to be round. Stars and planets were round. Tree trunks and plant stems were basically round. Many animals were round, and most of those that weren't became round when they curled up to go to sleep.

The more we thought about it, the more it seemed to Rick and me that animals preferred round shelters. As we looked through books and magazines in the library, we discovered that even most human shelters were round. Tipis, wigwams, wickiups, bark huts, sod huts, and igloos were all round. Many longhouses were more circular than rectangular. Even the shelters of African bushmen and Trobriand islanders were round. There were many variations, but the basic construction was almost always circular.

We found out that this was no accident. First of all, the circular design is one of the strongest structures there is. You can push on any part of a hoop and all the other parts will resist. You can jump on top of a dome and every part of the wall will take an equal amount of pressure. Every block in an igloo helps to support every other block. Circular structures are easier to build, more economical, and more solid. They require fewer materials. They shed water evenly and quickly. And they can take higher winds without being blown away.

I never talked to Stalking Wolf about these things. He seemed to take them for granted, the way an ant or bird might. What he did talk about was the spiritual power of the circle. He pointed to the roundness of dewdrops that gathered on leaves. He showed us the nests and eggs of birds. He pointed to the sun and moon and traced their arcing paths across the sky. He sat with us at night and showed us how all the stars turn in a circle around the star-that-does-not-move. We asked him why this was so, and he said it was because that was the way the Great Spirit had made it and it did not have to be understood to be appreciated.

Gradually Rick and I saw that the shelter was much more than it seemed to be. It became a symbol for the center of our existence. The fire in the middle was like the bright, fixed star of the Great Spirit. It was also like the fixed center within ourselves that gave rise to everything we would ever be or do. When we sat in our shelter, we could think of ourselves being at the center of our universe with everything revolving around us. Every day could be created freshly from inside ourselves, and the power of the center and the four directions would help us in everything we did. The cold wind from the north would help us to endure and grow strong. The winged of the east would help us to rise above our troubles and become farseeing and wise. The "mice people" of the south would help us to look within and pay attention to detail. And the wind from the west would bring rain to cleanse our spirits.

Many other realizations dawned on Rick and me as we grew in our love for wild nature. We dug up the earth as little as possible while making a shelter because the bare ground was a symbol for the nurturing power of the earth. The interior of the shelter was not only our protection, but the womb of creation. Its walls were like the sky. Even when we slept out under the stars, we could see how the earth and the sky were extensions of our shelter. This helped us to feel at home wherever we went. If we slept out, it was like being held in the arms of Earth Mother. If we slept in a shelter, it was like spending the night inside a small bubble of the Great Spirit.

Preliminaries

Emergency survival shelters give lots of warmth and protection, but they are very cramped. There is more to life than living like a caterpillar in a cocoon. The shelter described here is not a survival shelter in the strict sense of the word. It's more like a second home—a warmer, more luxurious place that you build when you can afford it. It takes more time and more planning. You'll probably be more picky about where you locate it and how you put it up. It's going to be warmer and lots roomier than anything you would put up for an emergency survival situation.

There are many types of shelters that fit into the advanced survival or "earth living" category. But I have decided to concentrate on one basic type. I call it the earthshelter, or earthlodge. It is a mixture of several different native American dwellings—the Ojibway wigwam, the Seneca bark hut or longhouse, and the Wampanog wetu. The dome shape gives it great strength and warmth. It can be made from a variety of materials on almost any terrain. It also has a sense of permanence that cannot be gotten from a tipi.

The tipi is a marvelous shelter, but it was used mainly by the nomadic tribes of the Plains, and it was meant to be moved when the buffalo moved or when the resources of one area were used up. The earthshelter is not meant to be moved. It is meant to be stationary so you can take the time to develop a balance with the landscape. It is meant to let you care for one small portion of the earth in a very intimate way.

Where to Build. Take plenty of time to choose a good building site. Look around. If you own a quiet acre somewhere, maybe you already have the perfect spot. Maybe you'll even want to put the earth-shelter on your own city or suburban property to remind you of the earth connection of your own home. There are lots of different uses for the earthshelter besides survival living. Aldo Leopold, the naturalist and university professor, lived in the city most of the time, but he kept a chicken coop in the country as a retreat. He recommended some kind of earthy "coop" like this for everyone.

Wherever you are, choose a high, dry spot that is well drained, well protected from wind and rain, and well exposed to the sun. Avoid dangers such as overhanging boughs and landslides or snowslides. Make sure the area offers easy access to fresh water and building materials. If you're in a survival situation without food, choose an area that offers a variety of wild edible plants and animals.

If your life is not on the line, look around with one eye focused on Mother Nature's needs. Many times I have found an excellent place to build a shelter but decided not to because I thought my presence there would hurt the landscape more than it would help. Sometimes this would be an area that had just been burned or scarred and needed more time to recover. Other times I could not exactly explain why; it just did not feel right.

Framework

I will explain three variations of the earthshelter design: the dome, the oval, and the many-sided. Each of these has its advantages. Domes and ovals are best in parts of the country where you can find strong saplings for the framework. The dome is a little simpler and more economical, but the oval is a little more convenient. The many-sided shelter works best in desert country or other areas where wood may be too brittle or where it makes more sense to use rocks, adobe, or stacked debris for the walls.

The Dome. The dome looks almost identical to the American Indian wigwam. It is made by marking a circle about ten feet in diameter (larger if need be) and sticking sturdy, ten-foot saplings into the ground around it about two feet apart. Opposite pairs of saplings are bent and lashed together to form a network of interlocking arches. The framework is then covered with bark, shingles, sod, mats, hides, or canvas.

Before you cut any saplings, slow yourself down and clear your mind. Take time to feel the pulse of the area and your connection to it. The object is not just to put up a shelter, but to establish a closer relationship with the earth. So ask the earth what it thinks. It will answer if your question is sincere. You will hear its echoes in your heart, welcoming you but asking you to please take care and to do as little damage as you can. You may get an inkling that it would not be right to cut the straightest and healthiest trees. If so, follow your feelings. Look for saplings that are crowded and have little chance of surviving. The land will not feel their loss so much, and they will be better for your shelter. Your inner voice may even tell you to use fallen branches instead of saplings. Stay open to that voice; it is the most important thing you can do.

Mark the circle either by pounding in a stake at the center and using a cord for the radius, or by pointing a long, straight stick from the center of the edge. After you have marked and cleared the circle, place two sets of saplings on the east and west sides. These will form the door frame. Face the doorway east or southeast so that it catches the warmth of the morning sun. If you also want a western entry to catch the afternoon sunlight, the "morning" entry can easily be closed off.

Bury the butt end of each sapling about eight to ten inches deep. The easiest way to do this is to pound a sturdy stake into the ground, work it around, and pull it out; then put in the sapling and tamp the dirt down with a stake. To make sure the saplings bow out when you bend them, angle the butt ends slightly inward.

Bend the ends of each doorway pair of saplings toward each other and twist or lash the ends tightly with cordage so there is at least a foot of overlap. (If you're a purist, you may want to make your own cordage from plant fibers or vines the way the Indians did. Rawhide cordage can also be used if it is pulled tight, though it will loosen if it gets wet (see *Cordage*, page 88).

When the east-west archway is done, connect the north-south saplings. Lash them together and bind them tightly to the first two. Then fill in the rest so you have an east-west archway with many sapling pairs crossing over at right angles. To strengthen the framework, lash all the saplings tightly at the intersections.

Now lash on the horizontal saplings. These don't have to be so thick and sturdy, but they should be strong enough to withstand the weight of the roof and the prevailing winds. Start at the bottom and work toward the top. If you plan to have a shingle roof, the distance between the rows of horizontal saplings should be about two-thirds the length of your shingle material. If you're using twelve-inch shingles, this means you'll lash on a ring of saplings every nine inches or so. If you're planning to cover the shelter with mats, a ring every eighteen inches will do. Make the doorway or doorways about three feet high. When the framework is done, the shelter should look like an airy beehive. When it's finally covered, the only open spot will be a smokehole about eighteen inches square in the very top.

An earthshelter about ten feet in diameter will normally be from five to seven feet high. You can make a larger dome, but it will probably require thicker saplings and they may not bend very easily. In this case, either thin the undersides or cut a line of narrow "V" notches along the underside. If necessary, heat them over a fire to soften them.

The Oval. The oval is very similar to the dome, just longer. It is modeled after the Indian longhouse, which typically had from two to four smokeholes and housed from four to eight families. Huge longhouses more than a hundred feet long and thirty feet wide were often used for ceremonial purposes.

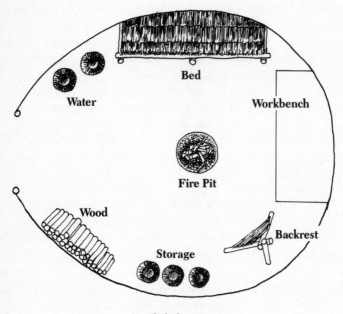

Earthshelter Interior

This oval adaptation can be made almost any size. Its main advantage over the dome is its added length and straighter sides. This makes for greater ease in moving around inside and easier placement of furniture such as sleeping platforms. The smokeholes in these shelters can also be put more toward one end, away from bedding and other household articles, and left open even during bad storms.

You don't have to actually measure the oval, just use your eye. After you've decided how long and wide you want it, sink a row of saplings two feet apart on either side of the shelter with a semicircle of saplings at either end. Connect and lash the side pairs first, then bend the end pairs over them. If the shelter is so long that the end pairs don't connect, lash more saplings between them. Finish the framework as before, by lashing several rows of horizontal saplings around the arches. Place these at distances that will be most convenient for shingling. If you plan to attach a bed to the wall, make sure you have a sturdy horizontal sapling a foot or more above the ground.

The Many-sided. The final variation is the many-sided earthshelter. It is basically an eight- or ten-sided shelter that gives plenty of room but doesn't require saplings. The roof can also be steeply angled, giving more room to stand up.

Start by marking out a circle (say, ten feet in diameter) and putting eight or ten stakes into the ground at equal distances around it. The more stakes you use, the more rounded the structure will be. Figure the distance between stakes by multiplying the diameter by three and dividing by the number of stakes. The number of sides depends on the size of the shelter and the weight of the roofing material. The bigger the shelter and the heavier the roof, the more sides it should have.

The support stakes in this case should be very strong because they will have to take the full weight of the roof. They should be two or more inches in diameter and sunk at least a foot into the ground. Be sure to consider this when you figure the height of the walls. For a shelter about ten feet in diameter, walls four to five feet high are ideal; this means you'll look for stakes that are at least five to six feet long.

Once the stakes are well secured, lash on crossbeams all the way around the outside. The beams will also take lots of weight, so make them strong, too. You can get by with lashing only, but it's safer to notch the uprights and crossbeams with "cabin notches" (like Lincoln logs) beforehand. This gives a snug fit and a lot more support. The notching can easily be done with a saw and an axe or chisel, or with tools made from natural materials (see *Stone and Bone*, page 101).

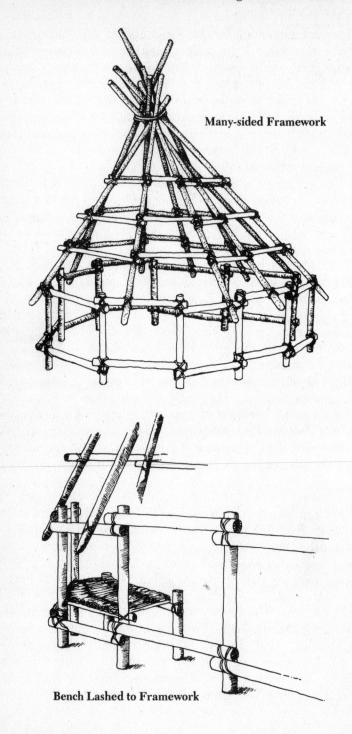

Many-sided Framework

Bench Lashed to Framework

The square lash is best for joining things at right angles. Start with a clove hitch over the end of one of the stakes, then wrap alternately several times around the upright and the crossbeam, forming a neat square pattern with the cordage. If you haven't notched the wood, tighten the lash by wrapping it around a few times in the center before tying it off with half hitches. (This is called frapping.) Make each lash as tight as you can without breaking the cordage. To increase the tension on the cordage, unwind it from a spool or stick that you can grab with your whole hand.

The roof on this shelter rises up like a tipi. It should be steep enough so that it will easily shed water and snow. For best results, start with three strong poles bound tightly at the top with a tripod lash. This "clove hitch" is followed by several turns around each pole and finished with a couple of half hitches. Lash the bottom end of each pole to the crossbeams with diagonal lashes to form a tripod. The diagonal lash starts out with a clove hitch, followed by two or three turns to secure it on the pole. Then the cordage is wrapped alternately around the intersecting poles to form an "X" pattern. Frap it in the middle as before. For a tighter wrap, unwind the cord from a stick.

When the tripod is secure, add more roof beams, one for each upright. If the structure is very large or you think the roof might be weak, first lash the ends of the roof beams to a strong center pole. Then add horizontal roof beams to support the shingles.

If you plan to shingle the walls, you can add horizontal braces there, too. But the many-sided shelter offers other alternatives. You can use rocks or logs by cementing the stacked layers with adobe (a muddy mix of dirt, water, and fibrous material). You can also make a "stacked debris" wall by forming a latticework of stakes around the main framework and filling it with insulating mateials. Grasses, reeds, leaves, mosses, and fir boughs are all excellent.

Square Lashing

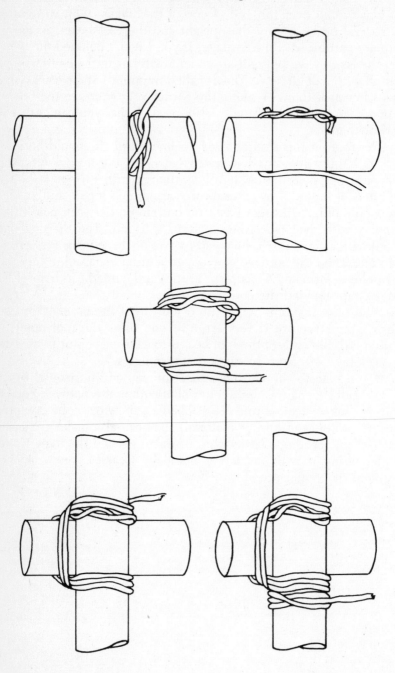

Tripod Lashing

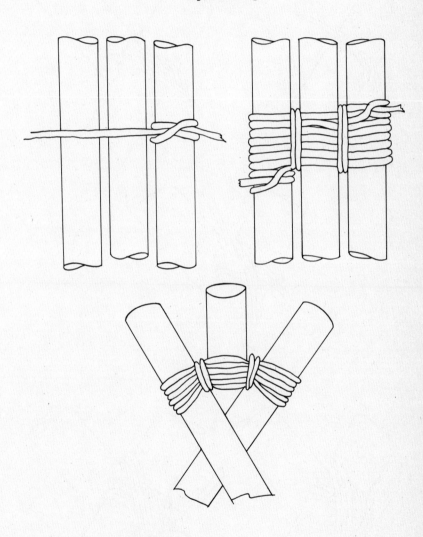

Diagonal Lashing

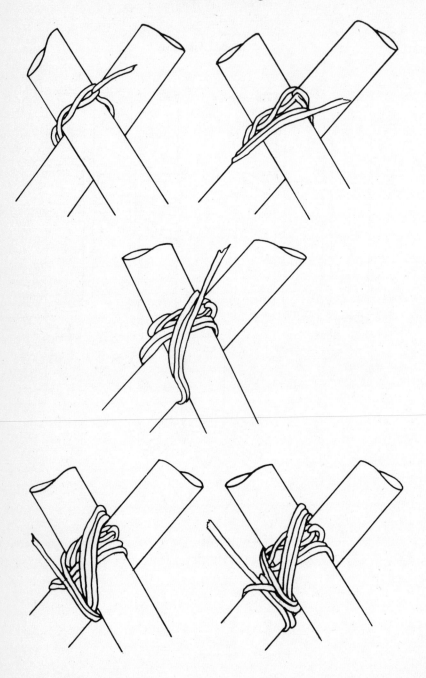

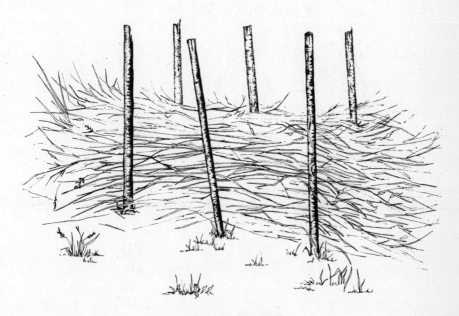

Stacked Debris Wall

Whichever design you choose, don't make the shelter so large or the roof so high that it's wasteful or impractical. In many respects, a small shelter is better than a large one. A ten-foot shelter can sleep up to four people.

**Cutaway View of
Many-sided Earthshelter**

Shingling

There are many ways to cover an earthshelter. Some people use animal hides, which are excellent but often difficult to get. If you do use hides, oil them well to keep them waterproof and pliable, and sew them together—ideally by first folding and interlocking the ends.

Grass thatching is another possibility, but I don't recommend it because it's so flammable. In my experience, the best natural roofing material is bark. Wood shingles are a good second, followed by tightly bound reed or cattail matting. You can always resort to canvas, which the Indians also used after the buffalo herds disappeared. But I would stay away from plastic sheeting and other synthetics, except for waterproofing.

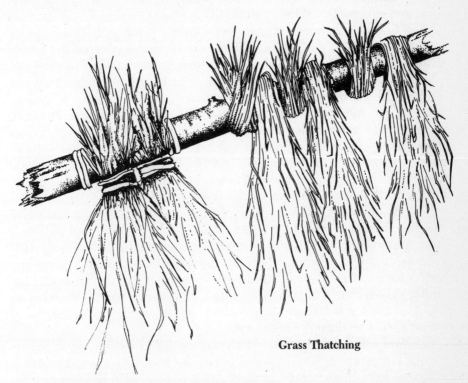

Grass Thatching

Bark Shingles. Birch bark is best for shingling because it's so strong and can be stripped off logs in huge chunks. Three- or four-foot lengths can even be sewn together and made into continuous mats that can be rolled up and spiraled around the shelter in overlapping rows. Smaller shingles can be made from cedar, pine, elm, and many other kinds of logs.

Shingles should always be placed starting with the row at the bottom and working toward the top. Each successive row of shingles should overlap the one below it by about one-third. To begin, notch or drill holes in both sides of each shingle a couple of inches down from the top edge. If you use notches, saw them at least an inch deep so that the shingles can also be overlapped on the sides when they're laced onto the beams. This helps to prevent wind and water from getting in underneath. Take your time. This roof is meant to last. You'll appreciate the extra effort when you're enjoying the dry shelter during a downpour.

Lash the shingles onto the horizontal beams by lacing over and under through the notches or holes, wrapping around the beam each time you go under. When you finish a row, the shingles should be packed tightly together or slightly overlapped. If you've built a dome or oval shelter, lay on the rows of shingles until you've covered the whole structure except for an eighteen-inch-square smokehole at the very top. Cover the smokehole with a slab of bark or matting that can be lifted and held into the wind with a long pole when you want to make a fire.

If you have a tipi-type roof, leave the top foot or so unshingled to allow for the smokehole. This can be fitted with a removable, cone-shaped cover made of rawhide or matting. Slit the cover so that you can open it up to form smoke flaps. These look a lot like coat collars, and they work almost the same way. They keep out wind and rain and create a vacuum that sucks out the smoke (see *Fire,* page 71).

Wood Shingles. If you have some simple tools, you can make shingles from any straight-grained wood. Cedar is best, and pine is a good second. Shingles should be cut into three-eighth- or half-inch slabs as large as you can reasonably make them. The best tool for this is the froe, a long wedge of steel fitted with a vertical handle. The froe splits off flat shakes from a log when hit with a hammer or the blunt end of an axe. Use dead wood whenever possible, not only for better splitting and burning, but to conserve resources.

Sod and Dirt Roofing. With a many-sided shelter, your beams may be strong enough to support a sod or dirt roof. This is most practical in drier areas, but a combination of materials can help to keep out the rain even in soggy country. First cover the roof with a latticework of poles, sticks, and brush; the tighter the better. Then chink the cracks and cover the entire roof with a layer of adobe (made by mixing dirt, water, and dried grasses or other fibrous material into a thick mush). Then add a layer of sod and mosses, or even a thick layer of brush and forest debris. The debris layer can help greatly to insulate the shelter. It can even be used on dome or oval shelters as long as the framework is strong enough to take the weight.

Mat Roofing. Natural matting makes excellent shelter cover. The best materials are cattail stalks and rushes, since they offer thick insulation and aren't too flammable. Cattail leaves, yucca leaves, and palm fronds are good, too. Grasses will do, but they have to be woven in bunches and they catch fire very easily. You can also use inner bark and other plant fibers, though it takes longer to prepare and weave them (see *Warp and Weft,* page 83). The best time to weave mats from vegetation is late summer, when the foliage is green but still supple.

Most of the Indians' roofing mats were four to five feet wide and eight to ten feet long. There are many ways to make them. One of the simplest is just to bind a row of cattail stalks or reeds together with cordage or strong plant fibers. If you're using cattails, hang a long row of them side-by-side on a string, or lay them flat on the ground. Alternate thick and thin ends so the edges will come out even. If you want a mat of double thickness, bend the cattails in half.

Once the cattails are laid out, tie a couple of half hitches or an overhand knot near the top of the first one. Wrap tightly around each of the next two cattails, then once around both together. Keep on in this way, wrapping once around two individual stalks and then binding both of them together. Repeat this process until you come to the end of the row. Then tie off the cordage, move down a few inches, and do another row of wrappings in the same way. The tighter and closer the wraps, the more waterproof the mat will be. To increase the thickness, bind in more cattail stalks wherever they're needed.

Another way to make roof matting is to start out the same way, but to use doubled cordage and bind each cattail with a simple overhand or square knot. Mats can also be made by sewing reeds or cattails together with a bone needle and cordage. This method was widely used by the Indians, but the resulting mats were quite airy and the cattails also tended to split more easily. Whichever method you choose, it's a good idea to tie wooden dowels on the ends. This helps to weight the mats down when they're on the roof, and also makes it easy to roll them up for storage.

Roofing mats can be sewn or tied together, but it's better to put them on like shingles. Start at the bottom of the shelter and work up. Either tie them directly to the shelter beams or weight them down with saplings and brush. On dome and oval shelters, arrange the mats so there's plenty of overlap to keep out the rain, and allow room for a smokehole at the top. On the tipi roof, spiral the matting around the cone to within about a foot of the top, overlapping as you go. Always arrange the mats with the reeds or cattails oriented vertically so they shed the rain (see page 48 illustration).

Double Walls

Once the outside shingling or matting is finished, you can increase the warmth of the shelter by putting some kind of watertight covering on the interior. Tying hides, blankets, or tightly woven matting to the interior beams makes a double wall with a thick layer of dead air space (see *Interior Mats*, page 47). In severe cold you can increase your warmth by packing this space with grasses, leaves, or other soft, fluffy material. The interior wall does several other things, too. It helps to channel leaks away from the sleeping area, cuts down on drafts, and adds a more finished look to the shelter interior.

If the interior wall comes clear to the floor, it creates an ideal chimney draft. As warm air rises inside the shelter, cold air is sucked up between the walls and helps to carry the smoke up and out. If you've ever been caught hacking and coughing in a smoke-filled shelter, you know what a blessing this can be!

Double-walled doors are also a good idea. In good weather, all you may need is a single reed mat or hide that can be rolled up or dropped down over the doorway. But in cold weather it's best to beef up the entrance. In case of severe storms you can even build a little alcove like the Eskimos did with their igloos. This cuts down on wind and snow and provides a place to take off wet clothing before you enter the living quarters.

Drainage and Ventilation

After the shelter is covered, dig a narrow trench around it to catch the rain. Be sure to place it directly below the outer walls so the runoff from the roof flows right into it. Also extend it on the downward side so that gravity carries the water safely away from your shelter. You might even want to pave the trench with bark so the water doesn't erode the sides of the trench.

Unless your shelter is covered with canvas or plastic, there's hardly any way to avoid leaks. If the rain doesn't seep in between the cattails, it will come in through the open smokehole. You can cut down on leaks by installing a waterproof inner wall as described above. Another is by smoothing out rough spots in the exposed interior beams so that water is channeled along them to the outer edges of the shelter. This is especially effective with the tipi-type roof because its beams are steeper. The best drip catchers of all are little stick spacers installed between the main poles and the inner wall. These naturally channel water toward the ground.

Double walls are a great help in ventilation. A good smokehole and smoke flap combination is also crucial to getting a good air flow. When the smoke flap is properly lifted into the wind, it protects against rain and forces the wind to flow faster as it rises up and over the flap. This fast-moving air sucks up cool, fresh air from below. If you have covered your earthshelter with mud, sod, or debris, always be sure to provide an extra opening so you have a constant supply of fresh air.

Furniture
Bedding can either be placed on the floor or on platforms built out from the sides of the shelter. I like these elevated platforms because they can also be used as benches and there's lots of floorspace underneath for storage. Many of the better wigwams had a whole ring of them around the inside.

These platforms are usually elevated one to two feet above the ground. Two forked sticks support the frame on one side, and a horizontal wall beam supports it on the other. The frame poles should be as sturdy as the main shelter poles. Secure the two end braces to the shelter beams with square lashes and place them into the "Y" of the uprights. Then put the long pole at right angles to these and lash the ends to the "Y" stakes.

To make bunkbeds, install vertical supports that go clear to the ceiling, then lash on a second platform above the first. The same procedure can be used for shelving.

Finish the platform by lashing short, thin saplings along the length of the frame. This can be done using a continuous over-under variation of the square lash. Pack these slats right up next to each other. When you're done, cover the frame with hides, mats, blankets, or some other type of bedding material.

Bedding. Bedding is traditionally a combination of hides and mats. Hides with the hair on are better because they offer more protection from the cold. With most hides, you can sleep right on top of the hair, but with deer hides it's better to turn the hair down and sleep on the skin side because they shed so easily. In any case, use plenty of insulation.

Interior Mats. Mats for bedding, blankets, floor coverings, and interior work surfaces are more useful if they're thin and flexible. For this reason, they are usually woven of soft materials. Weaving takes more time, but it's easy. And once you learn how, you can use the same techniques to make bags, baskets, and all kinds of clothing (see *Warp and Weft*, page 83).

Backrests. One of the most useful pieces of furniture is the backrest. Backrests not only ease aches and pains, but add a sense of comfort and even luxury to shelter living that comes in no other way. When made with care, they are as relaxing as lawn chairs.

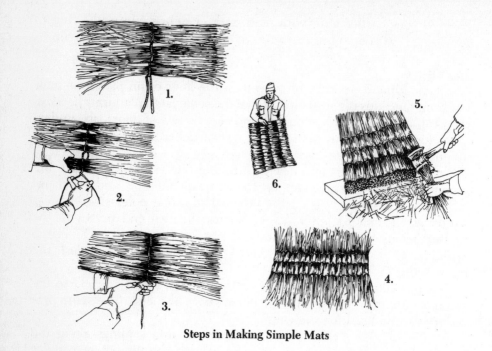

Steps in Making Simple Mats

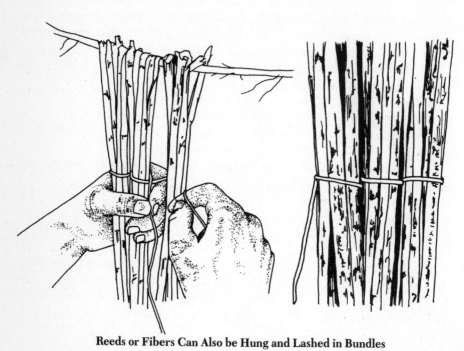

Reeds or Fibers Can Also be Hung and Lashed in Bundles

**Simple Backrest Using Sticks
Lashed to Tripod**

The backrest is usually supported on a tripod from four to five feet high. Join the sticks together at the top with a tripod lash and spread the "legs" to form a sturdy foundation. Another approach is to lash two legs of the tripod to one of the bunk uprights, using it as a foundation stake. If you do this, place the tripod so the backrest will face the warmth of the fire.

The backrest itself is ideally about five to six feet long and two-and-a-half to three feet wide. The extra width makes it more comfortable and helps it reflect heat from the fire. Traditionally the native Americans tapered their backrests from about three feet wide at the bottom to about two feet wide at the top.

To make a backrest, gather about 150 three-foot willow shoots from three-eighths to one-half inch in diameter. Peel off the bark by scraping with a knife at a right angle. (Do not carve!) Tie the rods in tight bundles and let them dry gradually for a week or so. When they are reasonably dry, unbind them and straighten the crooked ones (either with your fingers or between your teeth). Then string them together side-by-side with cordage.

There are two basic ways to do the stringing. The easiest is to loop the cordage and tie the sticks together in a series of square knots about two inches in from either side and down the middle. This will give three long rows of square knots equally spaced across the backrest. If you want to taper the backrest later on, tie each knot a little farther in as you go up toward the top. To keep it even, alternate thick and thin ends of the rods as you work.

A second method, more involved but more attractive, is to punch or drill three or four equally spaced sets of holes in the sticks and string them together. One way to do this is to stake out a pattern on the ground, tie strings in the proper places, and line up the dowel holes. Another way is to lay all the sticks side-by-side and run a knife or pencil along a straightedge to mark where the strings will go. The strings should be placed an equal distance apart with the outside ones a couple of inches in from the ends to prevent splitting. Make the holes either with a thin, sharp awl or with a drill. Then lace the cords through, securing them with knots on either end. Finish the backrest by lacing a loop of rawhide or other strong cordage around one of the upper rods so that it can be hung over the top of the tripod.

**Traditional Native American Backrest
Made by Lacing Thin Dowels Together**

Backrests were the Indians' most important piece of furniture. Most people painted lavish designs on both the backrests and the tripod stakes. Decorate these any way you want to liven up the shelter interior.

To make the backrest more useful and comfortable, face it toward the fire. In this way the heat will be reflected onto your body so you're warmed from both front and back. A second trick is to stuff insulation in the area between the backrest and the tripod. A third approach is to put matting, hides, or blankets over the top of the backrest. This cuts down on drafts and softens the sitting area. Even without these added comforts, the rods will bend and conform to the shape of your body, making for very comfortable seating. The backrest can also be used outside, either with the tripod or by leaning it against a tree trunk.

Benches. A simple bench can be made by notching and lashing. For a sturdier piece of furniture, I would suggest one of the forms of dovetailing. The first and probably the sturdiest method is the triangular dovetail. If you want to attach a leg to a bench platform, start by cutting a triangular notch into the side of the platform. Do this by sawing two sides at forty-five-degree angles to the edge of the platform, with the two cuts about right angles to each other. Then chisel out the interior. Next, shave the end of the leg into a dovetail. Taper the sides, top, and bottom as shown. When you're sure the leg will fit snugly, smear the end of it with hot pitch or hide glue and pound it into the platform. It should be wedged in so tightly that you can't pull it out. (see pages 52–53).

The same kind of thing can be done by making a round, triangular, square, or rectangular hole. This takes more time without a saw, but it can be done entirely with a hammer and chisel if you're patient. Once the hole is made, shave the end of the leg down to fit, tapering it slightly toward the end; then glue it and pound it in tight. For added security, drill a hole into the leg from the edge of the platform and insert a wooden pin. If you don't have a brace and bit, this can be done with a bow drill, using sand as an abrasive to speed the process (see pages 54–55).

Shelter Logistics

Generally, it's best to have the fire in the middle of the shelter so it radiates evenly to all parts of the living space. I would recommend setting a large tripod around the fire—one with horizontal braces at several different levels that goes all the way to the ceiling. This has many uses. Among them are drying meat, smoking hides, cooking, and storing utensils. It can even help to support the roof.

Triangular Notch I

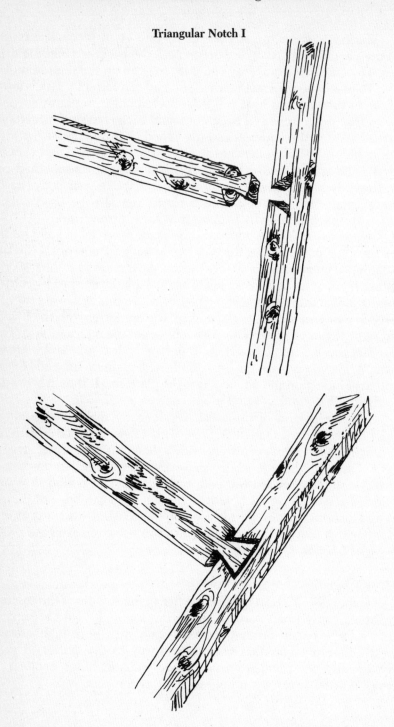

Triangular Notch II

The triangular notch makes a secure joint.

Making a Peg Insert

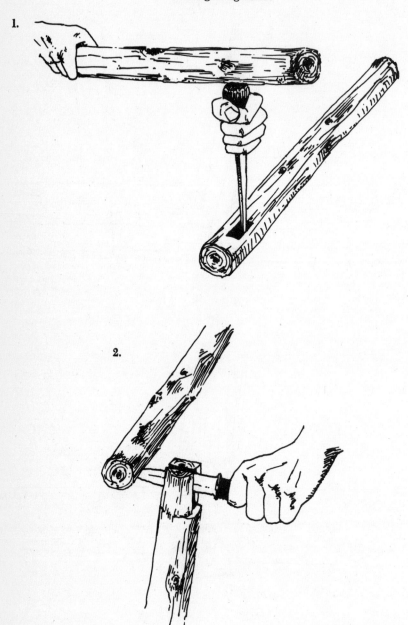

1.

2.

3.

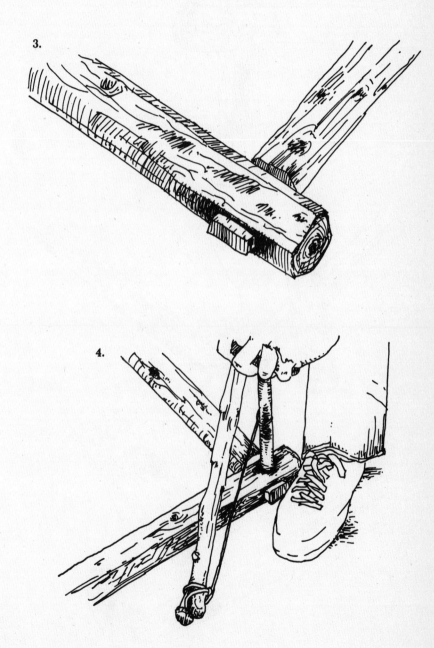

4.

Rectangular Notch With Peg Insert

5.

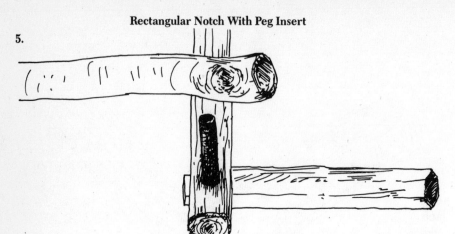

6.

7.

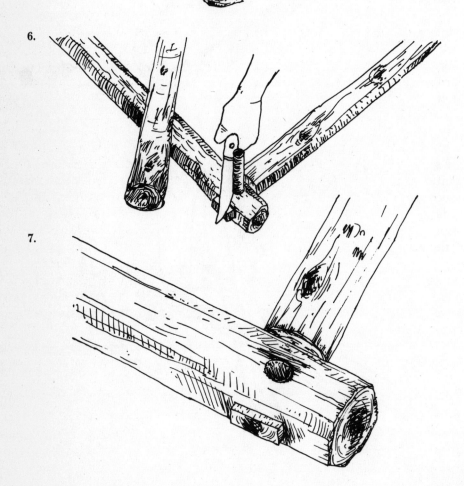

Water should be stored in a large earthen jug or other container somewhere near the entrance. Likewise, a good supply of firewood should be stacked just inside the entrance, with a smaller stack next to the fire and a longer term supply just outside the shelter. Tinder bundles and kindling should be kept close to the fire, or in some place where they will stay completely free from moisture.

Working, sleeping, and sitting platforms are best around the periphery of the shelter. The storage space beneath them can be used for rawhide boxes, baskets, pottery, bowls, and other durable items. Under one of the bunks you should also dig a pit where you can store meats, vegetables, or anything else that has to be kept in a cool, dark place. Nonperishable food items can be stored in covered clay pots or baskets beneath the platforms.

The Indians' shelters were lavishly furnished with floor mats, soft blankets, and beautiful hides draped over beds and backrests. Mats or bark should also be placed beneath storage items to keep them off the floor. In fact, most of the shelter interior can be covered as long as you burn wood that doesn't spark and the flooring is kept away from the fire itself.

Smudging. I recommend smudging the shelter interior periodically to freshen it and get rid of unwanted insects. You can do this by putting a few hot coals in a wooden bowl or large seashell and dropping a few sprigs of sage, sweet grass, or some cedar shavings on top of them. Mint and yarrow leaves are also good. These plants contain natural insect repellents, and they produce a beautifully scented smoke that can be fanned to all parts of the shelter. It's also a good idea to put cedar chips down on the ground to ward off insects, and to hang some fleabane in the rafters.

The Indians believed that smudging not only got rid of insects, but that it also warded off unwanted spirits and negative influences. Whatever your convictions, the shelter is your place of rest and renewal, and you should do all you can to make it a relaxing atmosphere. From any point of view, smudging is good medicine.

Appreciation

Most earthshelters do not cost much in the way of dollars and cents, but there is a price. The price is paid in the lives of the plants and animals and the care and dedication that go into the work. The investment of energy comes back very quickly, but not the lives that are sacrificed. It is sobering to realize how many lives it takes to make a single shelter. It is even more sobering when we realize how little most plants and animals need to survive themselves. While we are almost entirely dependent on the gifts that they give us, they are complete as they are.

These are good things to think about while making an earth-shelter or anything else. It is also good to feel a connection to the things we make, and to express our appreciation for the entities that go into them. It doesn't matter what our convictions are; we cannot live without taking other life. That simple realization helps to put our own life in perspective.

It is very satisfying to gaze at a well-made shelter. There is something pleasing about the way that it blends with the landscape. I feel a part of it, and I know it's a place where I can have my feet in the soil and feel my roots. In a house I often feel insulated from the ground and the atmosphere. I never feel this way in a shelter. As I look at the natural covering on my dome shelter, it reminds me of the back of the turtle, the Indians' mythical symbol for the surface of the earth. It reminds me of the animals everywhere that live in humble mounds and burrows. Inside my shelter, I know that I will be as comfortable as any squirrel or mouse in its nest or any owl on its favorite perch. To me, the shelter is the embodiment of living close to the earth.

Sitting inside on the floor or on a backrest the way Stalking Wolf used to love to do, I sometimes look up at the saplings arching over my head. I run my hands over their bark surfaces, as though over the backs of friends who have bowed down to protect me, and I know there is life in them still.

"Shelter trees," I say, "do not think that your sacrifice has gone unnoticed or unappreciated. The sap does not flow in your veins any-more, but your spirit is still strong. I can see it in your bark. I can feel it in your fiber. I can see it in your trunks that curve to meet each other like the ribs of the earth. Your first roots are gone, but now you will be bound in this circle and bring forth new leaves. Your leaves will be mine. They, too, will grow out of the earth, as all things do. My life will be nourished by yours. New warmth will rise beneath your arching backs. Where once you bent and swayed in the wind, now you will bend to protect this little patch of earth, this sacred circle that is now the center of my life."

3
EARTHBLOOD

We must listen
the water sings life
mother earth is
our natural delight
the water roars
and rises in spirit strength
the water tells us
we are one
                    John Trudell

Water meant a great deal to Stalking Wolf. In his lifetime he had gone without it often enough to know just how precious it was. He knew it as a friend that soothed his parched throat. He knew it as a cleanser that washed his clothes and cooking utensils. He knew it as the icy floor of the pond that supported his weight in the winter. He knew it as the bubbling liquid that cooked his food. But these were not the main reasons for Stalking Wolf's reverence. To him, water was the blood of Earth Mother, and it deserved a respect far above and beyond any practical use it might have.

In Stalking Wolf's view, water was magical. For more than eighty years, he had watched it gather in the sky, near the heart of the Great Spirit. He had seen it cast down to earth in many different forms. Sometimes it came as a soft mist, sometimes in pelting raindrops. Sometimes the Great Spirit sent thunder and lightning along with it to add power to its cleansing, and sometimes it fell in the form of lacy snowflakes to make a blanket for the season when everything sleeps.

To Stalking Wolf all the forms of water had the same spirit, and he loved to watch the cycle in all its stages. He watched the snow melt in the spring, when little droplets gathered and began to flow. He watched streams become torrents and saw how they rushed from mountain to sea through the veins of the earth. He saw how water reached out and gave life to everything it touched. Every animal and every plant was bathed in it. The earth itself was bathed in it. If the heart of Earth Mother ever stopped beating, or if her veins ever dried up, he knew the earth would surely die.

Whenever Stalking Wolf stepped into a shelter without water, he felt something was missing. A shelter did not really come alive until it had at least one earthen jug, dripping with dew, just inside the entrance. And a home was not really a home unless those who lived there had taken care of some other important water concerns. The shelter area must be well drained and also provide a safe means of bathing and waste disposal.

Drinking Water

As with emergency survival, the water source should be clean and pure. Ideally, it should be at least fifty yards from your shelter site. The water should be clear and fast-flowing, with a healthy assortment of plants and animals nearby. Lakes and ponds may be safe, but be very careful because they are much more likely to be stagnant or to pick up pollutants from runoff.

If you're living in an especially dry area, you can supplement your water supply by gathering dew from the ground in the early morning. You can also make a solar still. The still is nothing more than a large, square sheet of plastic with a rock in the middle that is draped over a hole. Depressed at a forty-five-degree angle, it causes gathering dewdrops on the underside of the plastic to slide down the edges and drop into a waiting container in the bottom of the hole. The liquid can be sucked up from the container through a length of plastic tubing.

Diverting. You may be able to set up a system so that flowing water is brought right to your shelter. Sometimes you can divert a little channel of water from a stream and direct it through your living space and back into the stream again. (Hollowed logs, bark slabs, or some kind of piping makes this easier.) If your shelter is located below the water source, you can siphon the water out with a hose. In this way you always have a fresh supply of running water and you don't have to worry about gathering and storing it. Another approach in rainy areas is to gather fresh rainwater off the roof through a system of natural gutters and downspouts that empty into waiting containers.

If you're in an area where water is scarce, find a low-lying area (ideally, a former streambed) that has green vegetation and other hopeful signs and dig until water begins to fill the hole. In promising areas, such holes can be used as wells to get already-filtered water by lowering containers on ropes. If you use a well, always keep it covered to prevent contamination.

Treating. If you have doubts about the purity of the water, be very cautious with it. Be especially careful in areas that might have been sprayed with chemicals, and don't take chances with giardia or other infectious diseases. Always treat questionable water by filtering and/or boiling it before drinking. Generally it's best to boil up to fifteen or twenty minutes to make sure you've gotten all the bacteria. If you don't want to take the time to do this, you can use chemical purifiers. There are also several small water filters on the market, including filter-type straws, that can easily be carried with you. The solar still is also good insurance.

Some of these purification techniques may seem very unnatural, but so is the state of our water. As a result of our carelessness, the blood of Mother Earth has been polluted to the point where there is hardly a pure water source anywhere. It's not worth taking chances. One bout with diarrhea can put you so far out of commission that your whole wilderness experience will be ruined.

Containers

Whether or not you set up a "piped" water supply, you're going to need some containers in your shelter. In a camping or other nonsurvival situation, it's easy to bring in pots, pans, and plastic bottles, and you may choose to do this. But if you want to keep your earthshelter natural, I would suggest making your containers.

The best water containers are made out of clay, and this process is outlined in a separate chapter (see *Earthenware*, Chapter 9). Containers can also be made from rawhide (see *Hide and Hair*, Chapter 8); by weaving watertight baskets (see *Warp and Weft*, Chapter 5); by filling a wet rawhide pouch with sand and allowing it to dry; or from animal stomachs and bladders.

A wide variety of useful containers can also be made from wood, through a process of coal burning and gouging. Using this method, you can quickly manufacture water storage pots, cups, plates, bowls, ladles, spoons, and many other household items.

To make a wooden container, first find a proper-sized piece of wood. With tongs or sticks, drop a few coals onto the area you want to hollow out. (Simple tongs can be made by bending a sapling, shaving the middle down for flexibility, and tying a piece of cordage between the arms.) Blow gently on the coals until they have burned down. Blowing through a tube or reed helps to direct the air. But don't blow too hard, or the wood may get so hot that it starts to crack. When the coals have burned down, work out the charred part with a rock or scraper (see *Stone and Bone*, page 101). Then add more coals and continue until the container is the shape you want it.

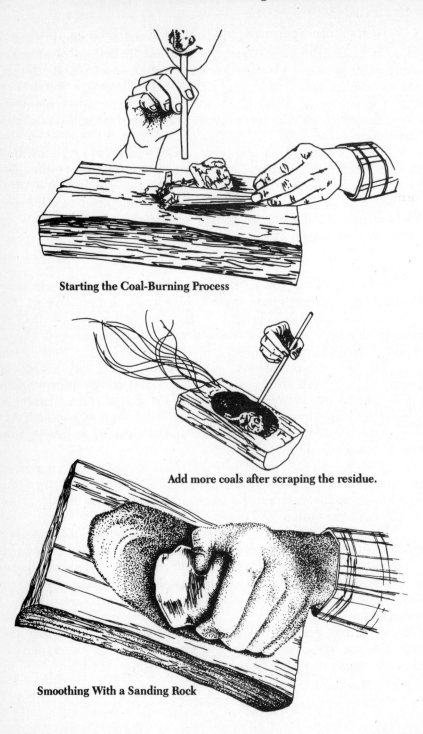

Starting the Coal-Burning Process

Add more coals after scraping the residue.

Smoothing With a Sanding Rock

The best tool for carving is the crooked knife. It has a J-shaped blade that gouges into the wood as you pull it toward you. You can make all sorts of containers with it, from cups and bowls to water buckets and grain grinders. Modern crooked knives are forged from steel. But almost every modern tool has its natural counterpart. The Indians made their carving tools from the front teeth of beavers and other gnawing animals, or from slightly curved flint blades bound to wooden handles. These were more like our modern hand adzes. Such tools were made in many sizes. When combined with coal burning and a lot of patience, they could be used to hollow out vessels as large as canoes. (For more information on these tools and their many uses, see *Stone and Bone*, page 101).

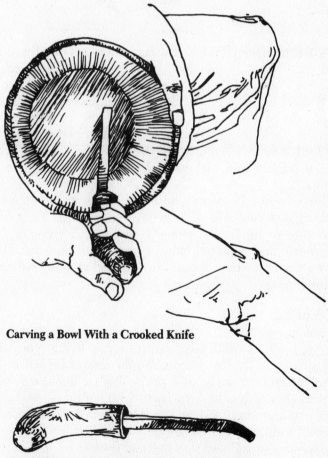

Carving a Bowl With a Crooked Knife

Crooked knives can be made of steel, bone, or rock.

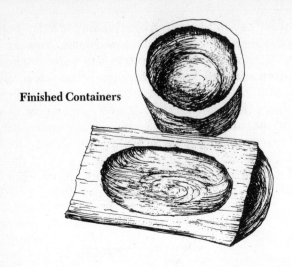

Finished Containers

It's best to make your first few containers out of softwoods such as cedar and pine. In time you may want to try harder woods such as walnut, hickory, ash, and oak. These are harder to work, but they last longer and don't crack so easily. Softwood containers often leak a little until the wood binds up. If this happens, pound some twisted cordage fibers into the cracks to speed the process. If you're not going to use the container for hot water, you can seal the inside by rubbing in a layer of hot pitch, or pitch mixed with wood ashes, to make a strong "survival epoxy resin."

Always keep a lid or flap on top of the water container to help keep it clean and cool and to prevent evaporation. To serve liquids without having to lift hot or heavy containers, I recommend ladles. These are easily made from chunks of wood with branches attached to them, using a combination of coal burning and gouging with a crooked knife or small adze.

Waste Disposal

Even in emergency survival, you have to be careful about not polluting your water. With longer-term earth living, you have to be even more careful. Human wastes should always be left at least fifty yards from the water supply, and preferably farther if it's convenient. The simplest approach is to dig a narrow pit, or privy. Make it about a foot wide, two feet long, and two or three feet deep—larger for more people. Each time you use the privy, shovel in a little dirt so that the wastes mix with the soil and don't attract flies.

For a seat, you can make a box platform with a hole in it, set two poles with a cross beam over the trench, or lash a beam between two trees. If you want to get fancy, you can even make a small shelter over the pit to protect you from the weather.

Also think about the water supply when bathing, washing dishes, cleaning animals, discarding garbage, and making tools. The earth will transform a lot of bacterial wastes if you give it a chance. You can do this by using biodegradable soaps, such as those from certain wild plants. (see *Utilitarian Plants*, Appendix). You can dig separate garbage pits to throw away things that are biodegradable. Better yet, you can return them to the earth where they will serve as compost.

Bathing

There are lots of ways to stay clean, from simple sponge baths to full immersion in a lake or stream. If your shelter is near a pond or lake, I do not recommend bathing in it unless it's very large and the water is flowing. It's much better to take some water back to the shelter and sponge off once in a while with a cloth or a wad of soft fibers. If you have a large tub, you can probably put enough warm water into it for a pleasant bath.

Streams and rivers are best for natural baths (with biodegradable soap). If they are too boggy, dig out a bathing spot. If they are too shallow, you may be able to build a levee that funnels the current through a narrower channel and increases its depth. The only thing I don't recommend is dams. If you make a dam or a levee, open it up periodically to let the silt wash downstream. Most dams impoverish thousands of acres of farmlands by locking up soil nutrients with the silt.

You'll find, as you divert or dig out water channels, that the effects of your actions are far-reaching. If you divert a stream to form a pond, for instance, you may attract more mosquitoes. You may also attract swallows and snakes and maybe even flood some animals out. As with other aspects of earth living, you will learn through experience what effects your actions have on the environment.

The effects aren't always negative, by any means. In the Pine Barrens camp where I teach most of my classes, there used to be just a narrow stream that bubbled up from the aquifer and ran through a bog. There were very few fish or turtles. A few years ago we dug it out to make a swimming hole. We cleared out all the old logs and a lot of the mud so that the stream flowed into a little pond with a clean, finely graveled bottom.

At first the place looked like a mess, and I got some complaints about it. But in time some interesting things began to happen. First of all, the new pond provided the only practical place for large numbers of people to bathe, and my students loved it. Everyone could get clean quickly, and everyone could experience the same kind of enjoyment that Stalking Wolf did when he laid back and let himself float with the current. We also made two sweat lodges nearby so that people could add a spiritual element to their cleansing (see *Sweat Lodge*, below).

But there were other spin-offs, too. Now we find that there are lots more fish, frogs, and turtles living in the stream. There are more mink, weasels, and raccoons, too. Deer, which have circled through the swamp for hundreds of years, used to avoid that part of the stream because they would sink up to their bellies in mud. Now they come down to the water's edge with no hesitation.

The benefits, in other words, were not only to us, but to other animals and even to plants. The disturbance caused some ugliness and discord at first, but as new plants grew and old ones rearranged themselves, eventually it took on a new balance and beauty all its own. It was a lesson in caretaking. It taught many people that it's possible not only to avoid harm to the environment, but even to do it some good.

Sweat Lodge. Another way to bathe is with a sweat lodge. Lodges were used by almost all native American tribes for a variety of cleansing and curing purposes, and were usually considered sacred. But for bathing you can make an unblessed lodge using the same techniques as for the domed earthshelter. Just make it smaller (say, six feet in diameter) and cover it completely, except for the entrance, to prevent hot steam from escaping.

Steam is created by pouring handfuls of water on red-hot rocks that are placed in a central pit. The pit should be large enough to hold about a dozen grapefruit-sized rocks. These should be heated in a roaring fire for at least an hour before you transport them to the lodge. If you use a bucket or other commerical container to hold the rocks, make sure the paint or galvanizing has been burned off first.

Don't stay in the lodge more than twenty minutes or so at a stretch. Also don't wear any metal watches or jewelry that might heat up and burn your skin. Most important, stay away from the rocks themselves, and keep anything flammable away from them. If it gets smoky inside, open up the door flap (usually a blanket or mat) to let in fresh air.

The sweat lodge is a beautiful place to think about the cleansing power of water. There in the dark you can create your own clouds. A blast of steam billows toward the top of the lodge. It curves and descends as though circling the earth. You can feel it spread its fingers over your back, face, arms, and legs. You can feel its heat drawing water from inside your body. It is the blood of the earth drawing out the poisons, cleansing your system. Every drop of water contains a little bit of the ocean. Even in the sweat lodge you can feel the pull of the tide. Its ancient rhythm tugs and purifies, reminding you of your origins and the unity of all life.

When you are done with a sweat, there is nothing better than to dash a bucket of cold water over your head, or jump into a cold stream. Suddenly you come alive again. The air seems fresh and clean. Your mind and body are crystal clear. The firewater has done its work. As the cold water washes over you like a violent storm, inside you can feel the thunder and lightning, and once again you know that the blood of Earth Mother has done its cleansing work.

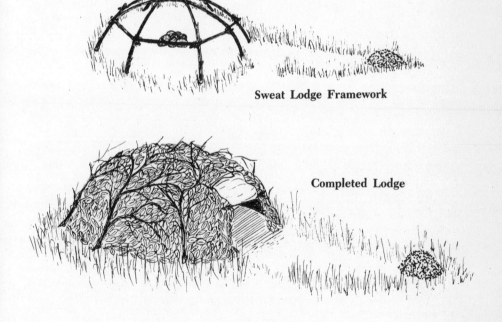

Sweat Lodge Framework

Completed Lodge

4
FIRE OF LIFE

Sunlight is at the heart of it all.
Sunlight is love. Love is a sacrifice,
especially when given to those who do
not understand . . .

                              Joseph Bruchac

When Stalking Wolf made a fire, he usually used the hand drill. The hand drill is a long, thin willow rod or mullein stalk that is twirled between the palms. Spinning and applying downward pressure at the same time, the idea is to get the butt end of the stalk to create enough friction against a fireboard to start it smoking and make a coal.

Rick and I had a difficult enough time learning the bow drill, which works on the same friction principle but is easier because you can apply a lot of pressure on the drill with one hand while spinning it with a bow held in the other. We had both tried the hand drill probably a hundred times without any luck. The blisters on our palms had long since turned to calluses, and we still couldn't make it work. Stalking Wolf came by once in a while, but he didn't say anything. He knew we had to learn through trial and error and that we needed to strengthen ourselves through repeated failure.

Eventually Rick and I both knew we were strong enough and had the right form, but we still weren't getting a good coal. We even teamed up, Rick sitting on one side of the fireboard and I on the other, and we would trade off so that when one of us had spun our way to the bottom of the stalk, the other would start out at the top to keep it going without letting the fireboard cool down. A couple of times we got a coal and excitely lifted the tinder to blow on it, but by the time we got it into the air it would go out. When Stalking Wolf came by, we were pretty frustrated. We told him we thought we had done everything right and asked him why it didn't work.

Stalking Wolf agreed that we were doing everything right, which only confused us more. Then he picked up the mullein stalk himself. He closed his eyes for a moment and placed it onto the fireboard. What he did then looked as easy as striking a match. One spin down the stalk, palms rubbing, stalk twirling, and the air filled with the smell of burning cedar. When the twirling stopped and the air cleared, Rick and I could see the little coal in the notch and the wisp of smoke rising up from it.

Stalking Wolf flicked the coal onto the tinder and bent down to pick it up. He wrapped the tinder carefully around the coal like a birdnest cradling a precious egg. Then he blew on it with a thin stream of air. He did everything we had done. The only difference I could see was that he seemed to be talking to the coal. He was coaxing and encouraging it with little gestures and expressions. There was no frenzy or excitement. It was more like a kind of affection that he was lavishing on the coal.

The coal grew and he continued to nurture it. He did this with the utmost care and patience, calculating just where and how to send the next breath and just how to compact the tinder so that the coal would grow and spread. He treated it like a little heart that was pulsating in the middle of an embryo. With each breath it expanded and glowed more brightly until the tinder was consumed in smoke. Finally the tinder burst into flames in his fingertips and he placed it under the kindling that had been propped up tipi-style inside the fire pit.

Rick and I watched as Stalking Wolf added more pieces of kindling to the tipi structure. It all looked so easy. The little fire flickered and licked upward. The flames began to dance and curl around the edges of the wood. As the heat radiated toward us, we asked Stalking Wolf once more what we had done wrong. He turned to us and said, "Did you ask for a fire?" Then he left.

Rick and I were confused. No, we had not asked for a fire, but we didn't see that it would make much difference. All the same, we tried it again, this time remembering the attitude Stalking Wolf had displayed when he made his fire. We opened our hearts to the Great Spirit. We thanked the fireboard and the mullein stalk and the tree that the tinder had come from, asking them all to bless us with a good coal. We thanked the sun for its energy and asked it to reward our efforts. Then, just to be sure, we thanked the plants and animals whose wood we had taken and we asked for their help, too.

I believe we did almost everything the same as before, but we did not feel the same about it. We felt more relaxed, and we had the distinct feeling that we were not working alone. We felt there were other entities in the shelter with us, peering over our shoulders, giving us the confidence we needed to make good use of our skills. This I believe gave our movements a fluidity that was more in tune with the materials we were working with, and we succeeded. We could not have done it if our skills had been lacking, but asking for fire had given us the edge we needed, and I cannot count the times it has worked since then.

This was a powerful lesson for Rick and me, and it stayed with us. In fact, it began to permeate almost everything we did. Asking for fire put it in a completely new perspective. We had thought we could create fire just by ourselves, but now we realized just how impossible that was. Our efforts alone could never make a fire. Sunlight and wood and plants and animals and water and many other entities went into it, too. Fire was not just a result of skill; it was a gift from the Creator.

This perspective is not easy to grasp when we know we can make a fire with a mere strike of a match or the flick of a lighter. But up until the last few hundred years, that is the attitude people had toward it. Fire was a sacred entity. It was a powerful purifier that could transform almost anything. It consumed the cold and darkness when the sun slipped below the horizon. It purified water and cooked food. In the sweat lodge it purged the body and warded off disease and despair. With time it even broke up and consumed rocks.

It's when we do without fire for two or three days that we really begin to appreciate it. It becomes precious beyond words. Even in nonsurvival situations we can feel this. Until that spark of life begins to flicker inside a cold, dark shelter, the shelter is dead. But once its heartbeat grows strong, the shelter comes alive with light and warmth that brings much inner peace and happiness into our lives.

As we cast our eyes into a fire, we can feel it not only warming our bodies but making our spirits glow. The effect is almost hypnotic. It's as though the coals held part of the secret to our existence. There is something in a fire that brings out our own inner glow and pulls us closer to the earth. Just as fire is the center of the earth, it is also the center of attention, the center of the lodge, and the center of the sacred circle in our hearts.

Fire Pit and Draft

There are several things you can do to enhance the warmth and light of a fire. One is to dig a circular fire pit about two feet across and six to eight inches deep in the center of the shelter. The sides of the pit should slope gently inward, like the sides of a shallow bowl. If you ring the pit with rocks, this will help to hold the heat and prevent sparks from jumping out onto the rugs or matting. It's not necessary to line the pit itself, but putting rocks inside can also improve the pit's capacity to retain and reflect heat.

The draft system can be simple or elaborate, but you have to have fresh air coming in to feed the fire and a place for hot air and smoke to escape. The smoke hole in the roof, combined with air movement outside the shelter, tends to suck out the hot, smoky air that rises from the fire pit. But you often have to experiment to get the right adjustment on the smoke flap. For best results, attach a "telltale" of some kind to a stick on top of the shelter so you can determine the wind direction by looking up through the smokehole. Always keep the smoke flap lifted into the wind.

For good draft, fresh air has to be drawn in through the base of the shelter and directed toward the fire. If you have an airy shelter, you may not need extra draft holes. But if you want to direct the airflow, it's better to make four draft holes, one for each of the four directions, around the base of the shelter. These can be easily opened or shut with slabs of bark. In calm, warm weather you can leave all the holes open. In a blustery storm, you can plug all of them except the one facing into the wind.

An even better draft system can be made by digging six-inch trenches in the floor of the shelter, one for each of the four directions. These are started just outside the shelter and extended all the way to the fire pit. They are covered with bark or wood slabs (recessed a little to prevent bulges in the floor), then covered again with mats or hides.

If you want to get really elaborate, you can even put some kind of piping inside the trenches. Almost any wide, cylindrical material will do—even drainpiping or a series of interlocking tin cans with the ends cut out. You can also make natural piping from logs by splitting them in half, hollowing out the insides, and lashing the two halves back together.

For all of the trench methods, you can use wood slabs to adjust or stop the airflow. Once you get the hang of it, it's a little like sailing. Just looking up at the telltale from inside the shelter can give you a sense of how and when the trenches need to be opened to keep the fire roaring hot and strong.

Smoke can be a real bother, especially when you get a downdraft or something interfering with the airflow. The first thing is to make sure the smoke flap and draft trenches are properly adjusted. Also make sure you're using dry wood. If smoke is a problem, you can make a primitive chimney by wrapping some kind of fire-resistant material around the tripod that encircles the hearth. This will help to contain the smoke. You can make a fireplace with a chimney out of rocks and adobe, or even install a stovepipe.

Firemaking Tips

The best fire for general use is the tipi fire. Its cone shape helps to focus the flames, keeps the fire burning hot and strong, and sucks the smoke straight up instead of letting it spread out inside the shelter.

Tipi Fire Set in Fire Pit Lined With Rocks

The tipi fire is made with four different grades of fuel. First is the fine, fluffy tinder bundle that nourishes the coal or flame. This is made from soft material such as cedar bark, cattail down, or other dry plant fibers that are crumbled and buffed into an airy ball. Second is kindling, the tiny sticks up to pencil thickness that are propped against each other, tipi-style, in the bottom of the fire pit. Third is a layer of "squaw wood," sticks that are from thumb-to wrist-size in thickness. And finally there's the bulk firewood that's put on when the blaze is going strong.

As you gather your firewood, think not only about your own convenience, but also about how you can do the least harm to the landscape. Gather wood as though you were pruning an orchard. For a good, smokeless fire you need wood that is dry and not too heavy to carry back. So you don't just go out and chop down the biggest tree you can find; you go into an area that has lots of brush and small trees. You go to a place where you're likely to find dead, dry wood that you won't even have to cut.

As you work your way through this area, ask yourself how you can do some good. With this attitude, you may begin to see places where trees have died and are blocking off the sunlight from other trees. These can be taken down and broken into manageable chunks. You may also find dead branches that crowd out healthy ones. By taking these, you can sometimes make it better for the ones that are left.

On the other hand, you'll also see dead trees that are providing homes for ants and woodpeckers, or maybe an old snag that's all alone on a mountainside. Out of a concern beyond our own needs, you'll leave these alone. In other words, be deliberate and thoughtful in your harvesting. This is when the science of survival becomes an art form—when you begin to treat the landscape with a respect that flows from a deeper understanding and connection. This attitude can be carried over into almost every aspect of survival and earth living.

Firemaking can be a very delicate art, especially in wet weather. Tinder, kindling, and squaw wood must always be dry. For this reason, it's best to gather fuel from standing dead trees or from dead branches on living trees that are located in a place that's protected from storms and low-lying fogs. In damp weather, never get fuel from the ground; it absorbs too much moisture.

Take care to provide a dry foundation for the tinder, and set up the entire tipi structure before making the coal or striking the match or

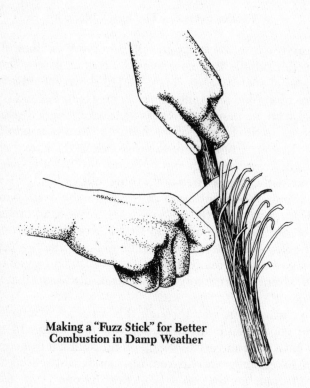

**Making a "Fuzz Stick" for Better
Combustion in Damp Weather**

lighter. There is nothing worse than to get a good flame that goes out because the kindling's wet or the squaw wood is too bulky. To help prevent this, lace the tipi structure with flammable materials such as dried grasses, leaves, and bits of kindling. These materials will help carry the flames on to the next layer of fuel.

If it's wet, have some "fuzz sticks" or pitch sticks ready. Fuzz sticks are made by carving little shavings into the edges of the thicker kindling pieces. This increases the surface area exposed to the flames and, if the wood is a little wet, helps to get down to the dry layers. Pitch sticks are kindling-sized pieces of wood made from the punky inner layers of decaying evergreen stumps. This wood is identified by the white, pitchy glaze that has formed on it. The pitch it contains is very flammable and burns for a long time with a bright, sputtering flame. Another good firestarter, if you can find it, is birch bark. This is especially effective because it flames up even when wet.

Pump-Drill Firestarting

The first book in this series, *Tom Brown's Field Guide to Wilderness Survival*, explains several methods of starting fires without matches. Among them are the bow drill, the mouth drill, and the hand drill. In all three methods, a wooden shaft is spun against a flat fireboard to create a glowing coal. The main difference lies in the method of holding and turning the drill.

An even more effective friction method is the pump drill. It's a little more complicated but much more convenient for shelter living, and it has lots of applications besides firemaking. The pump drill is a long, dowel-like shaft fitted with a horizontal wooden handle and a heavy, disk-shaped counterweight. The handle is drilled in the center to fit over the shaft, and the arms are suspended from the top of the shaft with cordage. Friction is produced by pushing down on the handle, thus spinning the shaft against the fireboard. The counterweight causes the shaft to spin back up automatically, and thus a fire can be started with a few easy strokes.

For firemaking, choose a long, straight shaft—preferably a durable one made of some hardwood such as ash, oak, or hickory. When the shaft is smoothed and straightened, it should be about two to two-and-a-half feet long and about one-half to three-fourths of an inch in diameter. The handle should be two to three inches wide, about sixteen inches long, and thick enough so it won't split under pressure. Shape the handle so your hands will fit comfortably on both sides, and drill a hole in the middle that will let the shaft spin completely free.

Cut the cordage long enough so that it reaches from the top of the shaft to both ends of the handle when the handle is at the bottom of the shaft. Cordage can be attached in one of two ways: either by drilling or punching tiny holes through the shaft and the handle ends, or by slightly splitting the ends. In either case, lace the cordage through the shaft and knot the ends under the handle to keep it from pulling through.

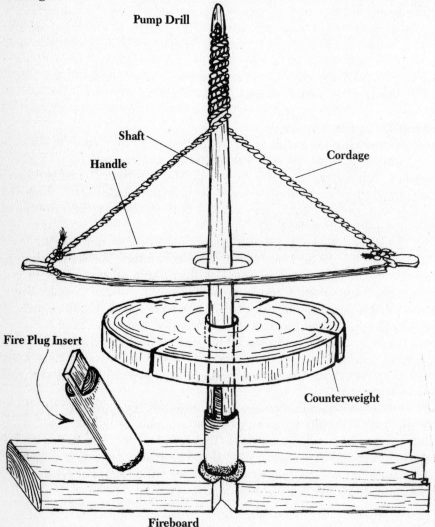

Pump Drill

Shaft

Handle

Cordage

Fire Plug Insert

Counterweight

Fireboard

If you want to make cordage for the drill, the best materials are reverse-wrapped sinew or finely braided rawhide strips that have been well buffed against a dull knife or sharp rock and treated with a light coating of neat's-foot oil. The rawhide should then be stretched, rolled back and forth on the thigh, and buffed once more over a twig or nail. Rawhide cordage like this can last for years. Strong, supple plant fibers such as dogbane or nettle can also be wrapped or braided into cordage for use with the pump drill (see *Cordage*, page 88).

The counterweight must be heavy enough so that the shaft puts pressure on the fireboard even on the upward stroke. For this reason, it's best to make a hefty hardwood disk about twelve inches in diameter and two inches thick. Drill a hole through the center of the disk just wide enough so the shaft can be pounded snugly through it. Secure the disk no more than three inches up from the bottom of the shaft with pitch or hide glue. For extra security, push pins through the shaft above and below it. If the shaft turns inside the disk, it can be secured with little wooden shims pounded up from below. For extra weight, you can also attach rounded rocks to opposite sides of the disk.

Once the counterweight is attached, cut a deep, thick notch into the end of the shaft—ideally by sawing or abrading with a rock. This notch receives the wooden plug that makes contact with the fireboard. The best plugs are made of medium-soft woods such as cedar, cottonwood, and willow, though in dry weather almost any kind of wood will do. The plug itself must be dry and of medium hardness. It should be two to three inches long and carved flat at the upper end so that it fits snugly into the notch at the end of the shaft. This joint should be wrapped tightly with sinew or strong cordage. The end of the plug that makes contact with the fireboard should be about three-fourths of an inch thick and tapered like a cigar.

Another alternative for the business end of the drill is a pithy shaft such as mullein or yucca. In this case, a tapered hardwood plug is first inserted onto the end of the hardwood shaft. This is then pounded into the pithy shaft so that the shaft splits slightly at the upper end and fits snug against the end of the hardwood shaft. It is then lashed on tightly with cordage.

Starting the Fire When the apparatus is ready, prepare the fireboard as you would for the bow drill or hand drill. The board should be of medium-hard wood, about half an inch thick, with a deep notch carved into the side to receive the hot dust from the twirling drill.

Before you cut the notch, first drill a little depression in the wood. To do this, turn the shaft in both hands, wrapping the cordage around it until the handle gets to the top. Then gouge a starting place into the fireboard, place the drill into it and push straight down with both hands. As you do this, the shaft and the counterweight will begin to spin. Let up on the handle about three-fourths of the way down the shaft and it will be lifted back up again by the spinning counterweight. Keep pumping with smooth, steady strokes and soon the drill will burn a slight depression into the board. Cut the notch (an "eighth of a pie") almost but not quite to the center of this depression.

Make a tinder ball of finely shredded, dry plant fibers (laced with cattail or thistle down, if available). Rub and buff it into a light, airy mass and place it on a piece of dry bark just beneath the notch. Then put the drill into the depression and begin to pump again. First concentrate on smoothness. When you've got the rhythm down, increase your speed until smoke pours out of the notch and you can see a fine wisp of smoke rising from the coal. Then flick the coal gently onto the tinder with a knife and blow it into flame.

Once you get the feel for the pump drill, it will rarely take more than ten downward strokes to get a good coal. This is vastly less effort than you need for other friction methods, but there can be problems. One of the most common is a wobbly counterweight. It should spin as smoothly as a top. If it doesn't, carve and sand it some more. Also make sure your drill and fireboard are completely dry. If you take time to hone all parts of the apparatus, it should work smoothly anytime you need it.

Drilling. You can also use the pump drill for drilling holes—in wood, bone, and even rock! For this you don't need such a big apparatus. Best is a one-foot hardwood shaft about the thickness of a pencil, fitted with a counterweight six inches in diameter and about an inch thick. The handle should be nine to twelve inches long. It can be operated with one hand, by grasping the handle with the drill between the index and middle fingers.

Drill bits can be made from hardwood, bone, stone, or steel (see *Stone and Bone*, page 101) and fitted into the shaft as described above. They should be made with a long, gradual taper on the point and an abrupt, wide tail on the other that fits into the notch in the shaft. The bits are lashed to the shaft much like arrowheads (See page 149). The drill itself must be harder than the substance you're drilling. If you want to drill through softwood, use a hardwood or bone drill. If you're drilling bone, use flint or some other sharp rock. If you're drilling rock, choose a rock bit that's harder than the rock you want to cut, add plenty of abrasive sand, and have patience.

Maintaining Fire

A fire inside a good, roomy shelter is much easier to maintain than a fire exposed to the weather. All you have to worry about is keeping it properly fed. The best woods for lots of heat and light are the softwoods, including pine, fir, hemlock, cedar, and the other members of the evergreen family. (Be careful with these, though, since the pines contain a poisonous resin and all of the evergreens produce a lot of dangerous sparks.) Medium hardwoods like cottonwood, poplar, aspen, willow, maple, sage, and alder are good for moderate heat and light. Best for long term burning are hardwoods like maple, ash, oak, hickory, and walnut. If you don't mind the extra smoke, you can also extend the life of the fire by burning wet or green woods. This is especially useful at night, when you may want to go for several hours without tending the fire.

Another approach at night is to mix wood chunks with hot coals and cover the firebed with a thin layer of dirt to slow down the combustion. Next morning, instead of having to start another pump-drill fire, you can just uncover the coals, add a few pieces of kindling, and puff a few times to get them blazing again.

The Longmatch

Storing Fire

The best way of storing fire that I know is the longmatch. It can keep a coal smoldering for hours or even days, which makes it especially useful at night or when you leave your shelter for long periods of time.

The longmatch is a little like a hot dog. It is made of a ventilated, combustible core surrounded by two thin layers of bark. The core is made by mixing punky, dry wood crumbs with tinder, peat, dried mosses, or other easily combustible materials. These are packed around a dowel-like wooden splint inside a tube of thin, fibrous bark such as cedar. It's like packing a pipe. You're trying to occlude the air enough so that it burns slowly but completely, so that the coal will gradually smolder its way through the entire length of the tube.

Once the core is well packed, remove the splint. This will leave an airhole all the way through the middle. Then wrap the core in another layer of fibers and finally cover it with an outer layer of bark. Secure the whole bundle by wrapping cordage around the outside.

To start the match, just put a healthy, marble-sized coal in the open end, blow on it a couple of times until it starts smoking, then crimp the end after the coal has burned down inside a ways. The coal will probably heat up the outer layer of bark, but it should not burn through.

The longmatch can be safely left hanging over the fire pit inside the shelter because it will smolder but not burst into flame. It can also be carried from one place to another. Just orient it so it faces into the wind and check on it once in a while to make sure it's still burning. If it burns down during your travels, start another fire and make a new match. When you want a fire, just open the match, dump the coal into a tinder bed, and blow it into flame.

Cooking Fires

Lots of cooking arrangements will work in the earthshelter. One approach is spit cooking over an open fire. This is simply done by setting up two "Y" stakes on either side of the pit and roasting the meat on a stick set between them. The tipi-shape fire works well for this kind of cooking, but you might also want to experiment with the log cabin fire, which spreads the heat more evenly over a wider area. For cups of coffee and other minor cooking needs, the star fire conserves energy the best. It is made by arranging pieces of wood like the spokes of a wheel, with the ends nearly touching in the middle over a small bed of coals.

The best cooking arrangement is to dig a shovel-shaped extension of the fire pit on the doorway side of the pit. This cooking trench should be about three inches deep and a foot or so in diameter (larger, depending on the size of the pit and the number of mouths you have to feed). When you want to cook, just rake a bunch of coals from the fire onto the platform and suspend the food above it.

If you're cooking stews, soups, or other liquids, you can suspend the pot or container from a wooden framework. The simplest is the one-pole spit. More sturdy still is a spit made with four "Y" stakes and two poles that cross each other at right angles. The stablest of all is the tripod framework.

If you dig a narrow trench, you may be able to place a pot or other container on the ground above the coals, or on a grate or grill placed over a circle of rocks around it. A lot of cooking can also be done right on top of the coals themselves.

Fire in Rocks

Finally, you can also cook with hot rocks. A single baseball-sized rock heated to red hot can be enough to bring a gallon of water to a boil. (Do not take these rocks from lakes or streambeds, as they may explode when heated.) Hot rocks can be used to heat and cook soups, stews, and any other kind of liquid. It's not even necessary to have a hard container. Rock boiling can be done equally well in containers of wood, clay, rawhide, and even internal organs such as stomachs and bladders.

Rock Heating. Rocks can also be used for heating different parts of the shelter. If you want warm feet, you can place warm rocks beneath your socks. If you want a toasty sleeping spot on a cold night, you can put warm rocks in or under your bedding like heating pads. You can even dig a trench, cover a bunch of hot rocks with dirt, and put your bedding on top of it. There's no limit to the different ways hot rocks can be used. Just be mindful of the fire they contain, and take care not to get burned.

5
WARP AND WEFT

I will take up the hollow heart of a reed,
bend it in water, dry it in the sun,
to a basket so supple and strong
that walking with it shouldered
the weather will pour in.

> ### Jane Hirshfield

Weaving is one of the simplest and most satisfying of all the earth skills. I can hardly begin to suggest the articles that can be made by wrapping splints and fibers together, or the beautiful variations that are possible using different materials and methods.

Indian baskets and trays of different shapes and sizes were used for winnowing, storing, cooking, carrying, and even trapping fish and holding water. They ranged in size from hardly bigger than a thimble to large enough to hold a human being, and the designs that were woven into them reflected the same joy and natural artistry that the Indians expressed in almost every aspect of their lives. Likewise, mats, bags, sashes, and clothing items were beautifully woven in a manner that was characteristic of each particular tribe.

Most of these things were intricately patterned and decorated. It was never enough just to make a utilitarian basket or pouch or sash. Almost always the maker felt compelled to weave his or her own personal prayers and medicine into the object as well. Thus the finished product was not only useful and colorful, but its interlocking fibers also contained part of the maker's personality. Like anything else, weaving was a way of communing with and honoring the Great Spirit. It was another way of giving thanks for the bounties of the earth and celebrating the great gift of life.

Weaving Materials

There is a wealth of potential weaving materials in nature. What you use will depend on what you want to make. If you want a crude survival basket or a quick mat, you can get by with grasses, reeds, branches, saplings, and other materials that require little or no preparation. If you want a fine, soft basket or garment, you can gather strips of fibrous bark from plants or trees. For watertight baskets, gather bundles of inner bark fibers and either work them into cordage (see page 88) or wrap them tightly into coils (see *Coiled Baskets*, page 95). Long, fibrous leaves such as yucca and cattail also make excellent weaving materials. They can be used in place of splints or vines for softer baskets, mats, and bags, or broken down and worked into cordage. Following is some more specific information on various weaving materials.

Splints. The strongest and least yielding of the weaving materials are splints. These thin, flexible branches or wooden strips are ideal for tough utility and pack baskets, door frames, seed beaters, and fish traps. They can be fashioned from vines such as wicker or grape, by splitting small branches or saplings, by stripping lengths of inner tree bark, or by pounding on log sections with a wooden mallet until the growth rings separate and peel off.

Ideal for tree splints are hardwood logs from oak, hickory, and ash. The logs should be cut in early spring, just after the sap has started to run. They should also be green and free of knots and branches. If the wood is already dead, soak it for one to three weeks before pounding.

As with hide tanning, this pounding process takes time. Try to get into a good rhythm. Lift the mallet or club in a way that is comfortable and conserves energy. I find that short, easy strokes are better than long, hard ones. It also feels more natural to push the mallet harder at the end of the stroke, finishing with a final snap of the wrist.

With repeated pounding, the log gradually begins shedding layers of wood that are just the right thickness for the splints. These wood layers are then scraped clean with a knife, sliced into thin strips of the desired width, and arranged to form the backbone of the basket.

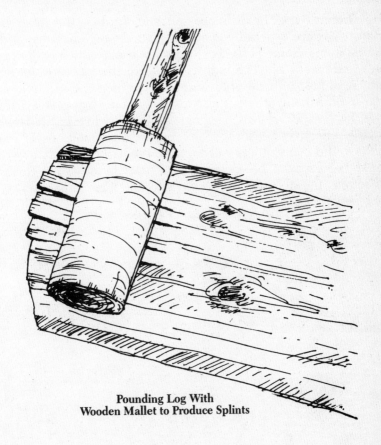

**Pounding Log With
Wooden Mallet to Produce Splints**

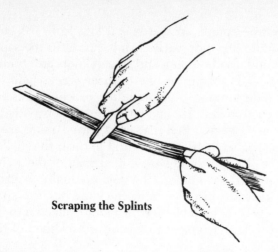

Scraping the Splints

Inner Bark. Being more soft and flexible, the inner barks of certain trees are especially good for mats, storage baskets, and rough clothing items. Some of the best trees are cedar, aspen, willow, cottonwood, and sagebrush. The Indians of the Northwest Coast, for example, made an almost unbelievable variety of clothing and utility items from cedar bark. Among others, these included mats, blankets, skirts, socks, shoes, hats, capes, shawls, leggings, sandals, and vests. Cedar was particularly effective because of its preservative and insect repellent qualities. But many others will do. All these barks are best for weaving if they've been soaked for several days or more in saltwater. This puffs up and softens the fibers. Soaking also gets the bark to peel off the wood more easily.

Peeling the Bark From the Wood

Softening the Fibers

Weaving

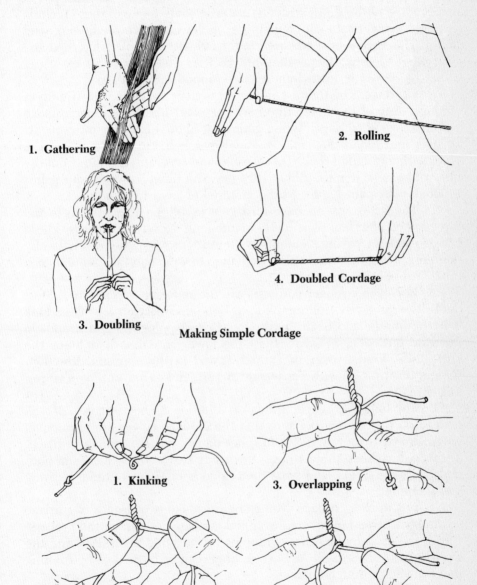

1. Gathering

2. Rolling

3. Doubling

4. Doubled Cordage

Making Simple Cordage

1. Kinking

3. Overlapping

2. Twisting

4. Holding

Reverse Wrap Method

Plant Fibers. The fibrous bark of certain herbaceous plants makes some of the finest weaving material of all. Some of these include dogbane, velvetleaf, stinging nettle, milkweed, fireweed, and wild hemp. This bark is best gathered in the fall, after the plant has died and it begins to loosen. Depending on the item you want to make, it can either be woven in loose bundles or wrapped into cordage.

Cordage. The tightest and finest weaves are done with strands of cordage. Any of the inner tree barks or plant fibers mentioned above will make good cordage. When gathering fibers from dry, pithy plants such as nettle and dogbane, it sometimes helps to gently pound the plant with a rounded rock. This crushes and separates the pith from the fiber ribbons. When the ribbons are dry, rub them between your palms or along your pant leg to rid the material of any excess pith.

The fastest way to make cordage is to roll the fiber ribbons in one direction on your leg, twisting them into long, loosely held cords. These can be used as weaving elements as they are, or strengthened further by doubling. Doubling causes the two halves to wrap naturally around each other in opposite directions.

For even stronger cordage, use the reverse wrap method. Start out in the same way, rolling the fiber bundle along your leg. Then kink it in the middle and begin twisting and overlapping the two strands in opposite directions. That is, twist the lower strand so it tightens the kink, then wrap it over the upper strand in the opposite direction. Repeat this process with the second strand, and so on, until you've run out of material.

For longer cordage, splice on additional fiber bundles. If you plan to do this, first make sure that the cordage halves are of unequal lengths so that you won't have to splice on two pieces in the same place, as this will weaken the cordage. Next, fray two to three inches of both ends, mix the fibers, and twist them together. Then continue twisting and wrapping.

For thicker cordage, you can either take two strands of cordage or double a strand of reverse-wrapped cordage and repeat the process. Finish the ends by knotting, wrapping, or weaving them back into the already-twisted cordage. Any excess fibers can be burned off by running them quickly through a flame.

Weaving In General

In most weaving there are two elements, the warp and the weft. The warp consists of all the vertical strands, while the weft consists of all

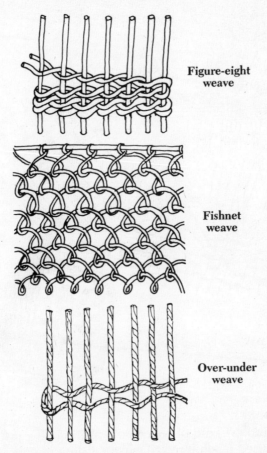

Figure-eight
weave

Fishnet
weave

Over-under
weave

the horizontal strands. (Sometimes the horizontal strands are also called "weavers.")

Whether it's a mat, basket, moccasin, or fish trap, the weaving process is much the same. The warp and weft are interwoven at right angles to each other. Generally with baskets the warp strands make up the framework or skeleton of the basket, while the more flexible weft strands are interwoven among them. With clothing items, warp and weft are more often of the same thickness and consistency.

There are three common types of weaving: plaiting, twining, and coiling. Following, I will describe each of these methods and explain how plaiting and twining can be adapted to making clothing, bags, mats, and other items. I will also include a short explanation of how to work with birch bark, since it can be used for many of the same purposes.

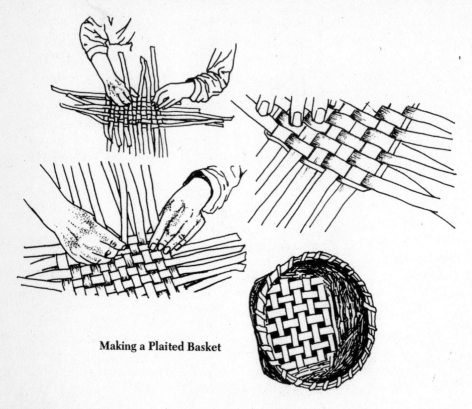

Making a Plaited Basket

Plaited Baskets

The simplest kind of weaving is plaiting. It's just like the over-under weave used by children to make rafts out of popsickle sticks. The form and function of the basket is dictated by the materials and the tightness of the weave.

Let's say you're making a plaited basket with splints. First decide whether you want the basket to be round or rectangular. If you want a round basket, arrange the warp splints like wheel spokes. Then, beginning in the center, weave around them in a circular fashion with the weft splints. If you want a rectangular basket, first make a "raft" of splints as you would with popsickle sticks. When you've formed the bottom of the basket, bend the sides upward and continue weaving in the weft strands. If the splints are brittle, or if you want a more angular basket, it helps to score them on the inside with a knife before bending them up.

Generally it's easier to make your first few baskets using wide, sturdy splints for the warp strands and thinner splints or plant fibers for the weft strands. Each time you add a new weaver, tuck the previous one behind the last splint and secure the new one between splints so that it's tightly held. With each new weaver, overlap at least two splints to keep them from pulling out.

Weave systematically all the way around the basket. As you go, the basket will have a tendency to bow outwards. You can prevent this by keeping the weft splints pulled tight. You can also cradle the basket between your legs to help it keep the shape you want. Tighten the weave as you work by using a bone or wooden awl to push the strands together.

To extend the sides of the basket, simply add more warp splints and keep on weaving. To do this, overlap the new splints with the old ones a few inches and stagger their lengths. This will prevent any weak seams. If you want your basket to be narrower toward the top, just push the splints inward and pack the weavers more tightly. Use your imagination to make the basket your own unique creation.

Finishing. Once you've tucked in the final weft, there should still be about two inches of warp splints sticking up. At this point, cut all the inside warp splints (every other one) flush with the top of the basket. These will be held in place by the last weft. Then cut off all the remaining warp splints about two inches above the top of the basket, bend them inward, and tuck them tightly between the second or third weavers. Soak them if necessary before bending.

Another way of finishing the rim is to cut all the warp splints off evenly and sandwich them between two splint rings around the top of the basket. These rings are lashed tightly to the warp by sewing around them with a thin, flexible splint or strand of plant fiber. Start this by tucking it underneath the two rings from the inside of the basket and bringing it over itself. Finish by tucking it back down behind one of the lower weft strands. No knots should be necessary, and the basket should hold together beautifully through the sheer strength and resilience of the splints.

There is a kind of magic to weaving that I can't help reflecting on each time I see a basket being made. Who would ever think that a tree could be transformed into a basket? You start out pounding on a log, taking the tree apart at the seams. You break it down and slice it into strips, then take those strips and weave the tree back together into a totally different entity. Yet part of the tree is still present in the basket. The same fibers that supported its limbs and gave it strength to withstand storms are now at work giving the basket the dynamic tension that holds it together for a different purpose.

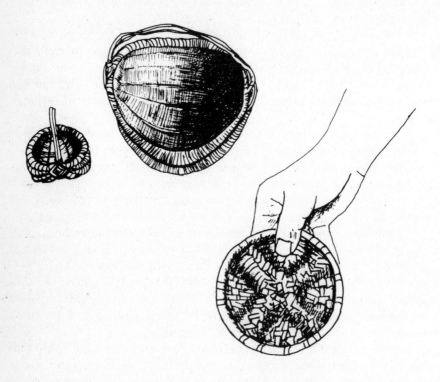

Twill Plaiting. After you've made a few baskets or mats with the plain plaiting technique, you might want to try twill plaiting. With this technique you can create interesting designs such as diamonds, diagonals, and herringbones. To do this, instead of weaving over and under each warp strand, you weave over and under two at a time—and you start one strand farther ahead or behind with each new weaver.

Wicker Plaiting. The uniqueness of wicker plaiting is in the material. Usually this is a vine such as wicker, grape, or honeysuckle. Once they're woven into the checkerboard pattern, they're a lot like flexible dowels or rods. They make some of the stiffest and most durable baskets and utility items around. Just be sure to soak the dry vines in warm water for half an hour before working with them, and keep them flexible by re-soaking whenever you feel it's needed.

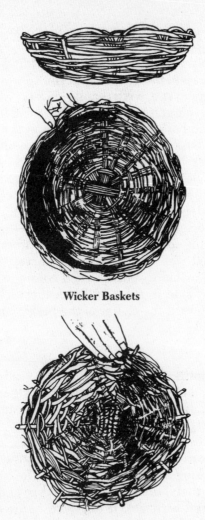

Wicker Baskets

To make a sturdy wicker basket with a handle, start with two vine hoops—one for the rim and the other for the handle and base. Strengthen the hoops by intertwining two or more strands of vine or wicker and lashing them at the ends.

To form the main basket framework, the horizontal hoop is placed within the vertical one and the two are lashed together at the intersections. This lashing is done by alternately wrapping around and between the crossed hoops, creating a diamond pattern or "godseye" that makes a little pocket to hold the ends of the warp strands. These strands are the "spokes" that form the bowl of the basket. The larger the godseye, the more spokes it can handle.

Start with only two or three spokes, holding the ends in place between the godseyes and the hoops, until you've woven in enough weft strands to hold them. (The weft can be wicker, vine, or any other material you like—even grass, bark, cordage, or reeds.) Gradually add more spokes until the basket takes on the shape you want.

To keep the basket even, I like to weave alternate strands on opposite sides. Otherwise it has a tendency to get lopsided. You can vary the shape and size of the basket by the length and placement of the spokes. Spokes can be added at any time, wherever the basket needs strengthening or widening.

When one piece runs out, tuck it back into the weave just below the last spoke—usually so the end can't be seen from the outside of the basket. Start the next weft strand in a similar way, first tucking it securely between two or more spokes. Then continue weaving over and under, working from side to side. To finish the wicker basket, simply weave each strand toward the center of the basket until it runs out and then tuck it in.

Twined Baskets

Twining is another useful weave. It's also called the figure-eight weave because each weft element contains two strands that cross over each other each time they are woven around opposite sides of the warp elements (See page 89 illustration).

Twining makes amazingly beautiful baskets, mats, and clothing items. There is a luxuriousness to this weave that's caused by the more complicated interlocking of the warp and weft. The twined weave is also very tight—so tight that if done with special care and the proper materials (such as prepared cordage), it can even be used to make baskets that hold water. Such baskets may leak a little at first. But as soon as the water permeates the fibers, they usually expand to make the weave watertight. For added insurance, you can then smear pitch on the inside as the Indians did. In general, twining is most often used to make pack baskets, seed beaters, fish traps, and other items that can take a lot of abuse.

There are some interesting variations to the twine weave. The simplest is the plain twine, or figure eight, as described above. Then there's the twill twine, in which each twist of the weft elements encloses two or more warp strands instead of one. There is also the wrap twine, in which one layer of splints is placed perpendicular to another and the two layers are wrapped together diagonally at the intersections with a flexible strand. I would suggest mastering the plain twine first and experimenting with the others as you develop your skill.

Coiled Baskets

The technique of coiling produces the most tightly woven baskets of all. In this process, the basket is made as a continuous spiral of rods or fiber bundles. Each rod or bundle is tightly wrapped and bound to the adjacent one by a fiber "thread" that is sewn around the coils.

There is a great variety of materials you can use, depending on the results you want. The softest, most flexible coiled baskets are made with bundles of grasses or pine needles. Harder, more durable baskets can be made by coiling long roots or supple willow shoots. The thickness of the basket is determined by the thickness of the bundles. For thin, flexible baskets, use thin bundles of grasses or needles. For thick, durable baskets you can use rods, either singly or in bundles.

Before you begin, make sure the material is soft and pliable by soaking it for half an hour in lukewarm water. Pine needles can be soaked for a couple of hours, or even wrapped in a damp towel for a day or two, as long as it's not so warm that they begin to mildew.

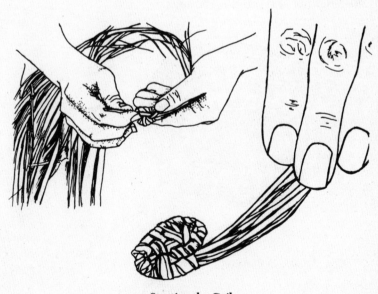

Starting the Coil

There are several ways to start a coil. If you're using willow shoots or some other kind of rod, it's often easiest to start spiraling from one end. If you're using pine needles or fiber bundles, it's easier to bend the bundle into a zigzag and stitch the zigzags together into a little platform. You can also tie several strands into a knot and start spiraling around the knot, or even wrap the first few bundles around a small piece of wood. Any method that forms a good base is fine.

The stitching should be done with a fine, flexible fiber such as a long grass, a strip of inner bark from a tree, or the strong, stringy fibers from a plant such as nettle, dogbane, velvetleaf, or yucca leaves. The bundles can be sewn together with a large commercial needle or with a bone awl or needle. Stitches are usually kept very tight, each one wrapping around the new bundle and then punching through part of the adjacent bundle to bind the two together. If you're using rods, each stitch should go around both the new rod and the one beside it, or around the new bundle of rods and one or two of the rods in the adjacent bundle.

Continue wrapping and stitching, splicing new bundles or rods onto the ends of the old ones. If you want a bowl-shaped basket, spiral outward and upward. If you want a cylindrical basket, form the flat base first (which can also be used as a mat or hotpad); then spiral the coils upward on the outside of this base. Vary the shape of the basket by varying the angle of the bundle lashings.

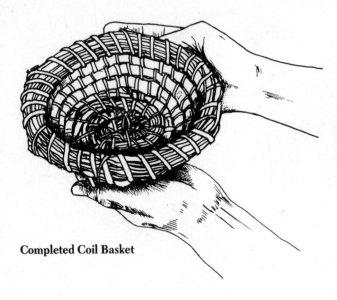

Completed Coil Basket

You can also vary the tightness of the basket with the lashings. If you want a loose weave, concentrate mainly on wrapping, and lash the bundles together every few wraps. If you want a tight weave, lash the fiber bundles together with every stitch, and stagger the stitches on each new row so that they fit neatly between the stitches of the previous one.

For watertight baskets it's best to start with material that's fairly dry but still pliable. This way it can be woven very tight. It will expand and tighten up even more when it gets wet. If the material is too green, it will shrink and loosen up as the water evaporates. This may cause the basket to leak, unless it's smeared on the inside with pitch or some other waterproofing substance.

Birch bark Containers

One of the best basket and utility materials is birch bark. It is strong, waterproof, has preservative properties. It also comes off in large sheets so that it doesn't require any weaving. The sheets are cut, bent, and bound like rawhide with thin rootlets or other utility "thread." With it, you can make all kinds of dishes, baskets, trays, and other vessels—even for cooking or holding water.

Paper birch is a good basket material, though black birch is more rugged and substantial. Look for a tree that's free of knots and branches. In most places where birch grows, you can find all the bark you need on fallen dead trees. This comes off more easily and you don't have to kill a living tree to get it. If you take bark from a living tree, don't girdle the tree or you may kill it. Carefully slice down vertically on both sides of the section you want and peel off the bark with your hands. With branches taken from live trees, soak the wood in water for an hour or so to loosen and make the bark more pliable.

If you take off the outer layer of birch bark you'll often reveal a peach-colored layer that is quite a bit cleaner. It's best to do this before you begin working the basket. If you prefer, leave the outer layer on, score designs into the bark, and peel off the outer bark within the designs.

Birch bark baskets and containers can be made almost any size and shape. Your imagination is your only limitation. Use spruce rootlets, grapevines, or cordage fibers to sew up the sides. Rootlets sewn in a spiral with a whip stitch along the top edges help reinforce the finished product. To avoid splitting the rim of the basket, stagger the stitches.

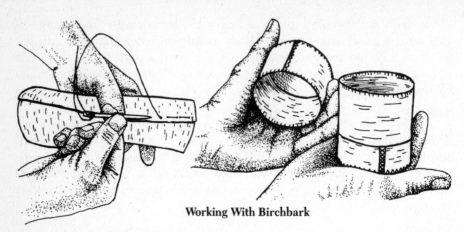

Working With Birchbark

Weaving Clothes, Bags, and Mats

Plaiting and twining are just as effective for weaving clothing as they are for baskets. Whether done on a loom or simply holding the materials in your fingers, the basic weaves are the same. The best materials for interior mats and clothing are the softer inner barks—particularly those that have been wrapped into cordage. (For information on roofing mats, see *Earthshelter*, page 29.)

For small things such as sandals and socks, you won't need much room. You can work right in your lap in the middle of your shelter. For larger things like blankets, robes, and interior mats, you can simplify the process and get a tighter weave by working on a makeshift loom. This may be as simple as a cord or a pole suspended between two trees, two vertical "Y" stakes fitted with a crossbeam, or four poles square-lashed together and set on the ground.

To start weaving, lay out the cordage or fiber strips for the warp strands parallel with each other, or tie them between two poles on a makeshift loom. The closeness of the strands will determine the tightness of the weave, and the number of warp fibers will give you the width of the article. If you're just beginning, it's helpful to vary the color or material of the warp and weft so you can easily tell them apart. This also creates a more interesting pattern.

Once the warp strands are set, begin weaving in the weft strands just as you did with the baskets, using one of the plaiting or twining weaves. When you come to the last warp strand, weave around it and start back the other way. Splice in new strands by fraying and interlocking the ends of both fibers, and then wrapping them into a single piece of cordage, or by tucking in the ends.

Garment weaving is simple once you get started. Even widening and thinning the material is not very difficult with a little practice. Wherever you want to widen the garment, just add more warp strands in the same way you added more splints for a wider basket. (Stagger the strands so the material stays strong.) To thin the garment at any point, just cut out or remove a fiber or two with each row. As you work, you'll get a better idea of how many fibers per row should be added or taken out to give the width and contours you want. To finish, pack three or four weft strands tightly together. Leave a couple of inches of warp to be frayed or tucked back into the weft.

If you're weaving a fine garment with cordage fibers, you won't have to treat the material after the weaving. With a more crude weave, such as with thick, unwrapped fiber bundles, you can increase the fluffiness of the garment by buffing the material after you're done.

Reflections on Weaving

To me, finely woven baskets, mats, and garments are beautiful symbols of the earth connection. Like a basket, the earth itself is a cradle of life that binds us all together and gives meaning to all we hold dear. Just as our own baskets help us collect, protect, and store our goods and even cook our foods, the earth holds all life's necessary ingredients, providing shelter, water, fire, and food to protect and nourish our bodies and spirits.

I often think of long-term survival as a kind of weaving, too—a kind of web that is spun with great care and concern and artistry. This is the way the Indians lived. Their culture and traditions were woven from nature, blending beautifully with the earth in a very flowing and graceful manner. Living this way, the native Americans saw great symbolism in their baskets and woven garments. In the strands that held these things together, they could see part of themselves, and a great many stories were told about the meaning of community and cooperation that was reflected in the warp and weft.

In the weave they could see how their own lives were intertwined and how the strength of the tribe depended on the fiber and orientation of each individual. In the tribal system there was no such thing as an unimportant person. Just as one strand made the difference between a basket that held water and one that leaked, the safety and well-being of the tribe often depended on the courage or quick thinking of a single individual.

In their weaving, the Indians could see the web of their own existence and more clearly understand how they were part of the spirit-that-moves-through-all-things. They could also see that their own tribal traditions were like the distinct patterns of the weave, bonding them together in a common purpose, and how even the tensions and pressures of each other's company could work to strengthen the fabric of the tribe. Most of all, they knew that no matter what the trials of life, they would never be alone. They would always have each other. They would always have their visions and their guardian spirits. And by accepting their individual roles and working in harmony with each other, they would weave the basket that held their collective destiny.

6
STONE AND BONE

Like the nails on a beast's paws, the old
tools were so much an extension of a man's
hand or an added appendage to his arm that
the resulting workmanship seemed to flow
directly from the body of the maker . . .
Eric Sloan, A Museum of Early American Tools

From the dawn of man, stone and bone formed the primary tools that allowed the human race to become all that it is. With vision and imagination, people down through the ages shaped these simple materials into extensions of the arms and hands that could do specialized work. In the beginning, most of these tools were devised to accomplish tasks necessary for survival. Unworked stone and bone were first used as crude weapons such as clubs and throwing rocks. Later they were crafted into finer tools such as knives, awls, scrapers, needles and arrowheads.

When we see such tools in museums today, we are often likely to think of them as primitive. But their appearance is deceptive. Many of these tools and weapons are the result of thousands of years of evolution. A better way to look at them would be as the finely honed survival tools that have paved the way for modern civilization.

The simple materials of stone and bone, so overlooked by most of us today, were seen as precious commodities not so long ago. The Indians habitually kept their eyes out for useful stones wherever they went, and they religiously collected the bones of animals that they could make into tools, weapons, and ornaments. I hope that in reading this chapter you will begin to look at stone and bone in a new light.

Stonework

There are many different kinds of stone. Each has its own properties and uses. Sedimentary rocks such as sandstone and limestone, which are formed in layers near the earth's surface, are composed of grains that have been compacted rather loosely under low pressures. Because of this, they make excellent grinding, sanding, and polishing tools, and they are also easily shaped by harder rocks to form grooves

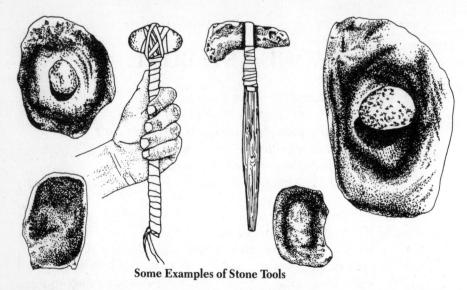

Some Examples of Stone Tools

and even bowl-shaped depressions (see *Pecking,* page 104), but they're too blunt and soft to make good, sharp-bladed tools.

Metamorphic rocks such as chert and flint are basically sedimentary rocks that have been exposed to intense heat and pressure at greater depths in the earth's crust. This tends to cement the grains together. They become packed and compressed so tightly that they begin to melt into each other. This makes a much denser rock with finer crystals. Because of their almost glassy nature, they cleave more sharply and make better blades. Some of the best cutting tools were made from these rocks.

Igneous rocks start out as molten material and cool at varying rates. The slower the rock cools, the more time its crystals have to grow. Rocks such as granite that cool slowly are composed of relatively large crystals. For this reason, granite makes a good, hard pounding tool but not a very sharp blade. On the other hand, igneous rocks like obsidian that have cooled almost instantaneously have crystal structures as fine as glass. Like chert and flint, they cleave very smoothly and make excellent knife blades, arrowheads, and other sharp but fragile tools.

Generally, the finer the crystal structure, the sharper the tool. This does not mean that you can't make a knife from granite, or that you can only use sandstone as an abrasive and pounder. The important thing is relative hardness. Generally, if you pound or saw anything with a harder substance, eventually it will give.

The first step in making stone tools is to get acquainted with different kinds of rocks. When you're in the wilderness, be aware of the variety of rocks that you see. Look for ones that are about the right size and shape for different kinds of tools. Pick them up. Test their weight and feel. Hit them with other rocks to see how they cleave and what their crystal structure is like. In other words, "talk" to different rocks and get to know them. In time you will be able to recognize different kinds at a glance and know just what you can use them for.

Sanders and Pounders

The simplest tools are the sanders and pounders. These rocks are used to shape other rocks and tools. The best sanding rocks are uncut pieces of sandstone and limestone, though other rocks can also be used. It's a good idea to collect several different sizes, grades and shapes. Sanding rocks can be used for all kinds of abrading, from smoothing bows and arrow shafts to cutting through wood and bone. Generally these rocks don't have to be worked at all. They are simply collected and used as they are.

The best pounders are medium-grained rocks such as fine granite and basalt that won't crumble or splinter on impact. These are often used for things like hammers, corn grinders, and meat mashers, though often you can find a rock that will also make a good, sharp hatchet or tomahawk. In fact, the Indians often made heavy cutting tools from medium-grained rocks like granite and basalt because they were heavy, didn't break easily, and could be sharpened simply by abrading the edges.

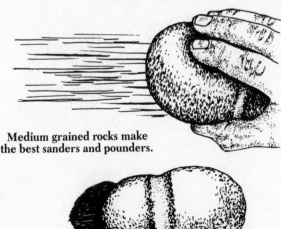

Medium grained rocks make the best sanders and pounders.

Pecking. Sometimes the pounders and grinders are used just as they are. Often, though, you have to attach them to handles to get better leverage. The best way to get a snug fit for the handle is to form a shallow groove all the way around the rock. This is done through a process called pecking—that is, hitting the stone a series of light, glancing blows with a harder stone. This gradually crushes and abrades the crystals to form the desired groove. The tool is then fitted and lashed onto a wooden handle with rawhide or some other strong utility cord (see *Hafting Tools*, page 114).

Pecking takes a lot of time. It is not something you should do when you're impatient or in a hurry. It's better to work on projects like this at odd moments—around the fire, while talking with friends, or while waiting for a hide to smoke. In other words, use it as a time filler. Hold the pecking rock in one hand and the tool in the other and just keep striking those soft, glancing blows. As you hit, you'll see the powder in the air and the broken crystals begin to fall, and soon a depression will form. You can speed the process by wetting the surface of the rock. Just keep turning and hitting the rock, being as patient and persistent as flowing water.

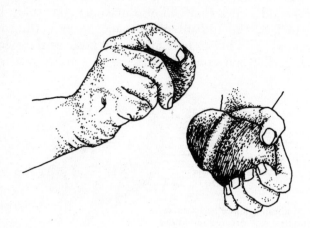

Pecking a Groove for a Handle

Sharp-Bladed Tools

For tools like knives, scrapers, skinners, hatchets, and arrowheads, you generally need sharper edges. Just how sharp depends on the tool. Edges vary from almost blunt for hide scraping to razor sharp for an all-around utility knife or arrowhead.

Sometimes fairly sharp-bladed tools can be made by abrading fine-grained igneous rock such as basalt. However, the sharpest instruments are made from rocks like chert, flint, jasper, quartz, and obsidian. Of these, chert nodules are probably the most common, though you can even use beer bottles for practice. If you use chert nodules, it's best to harden them first by burying the nodules six inches underground beneath a blazing fire for four or five hours and then letting them cool.

Direct Percussion. With a glassy or semi-glassy rock, you can produce a sharp edge by hitting it a hard blow with another rock. This is called direct percussion. It's used both to break up big hunks of core rock and for shaping the large flakes, or "blanks," that later are honed into fine-bladed tools.

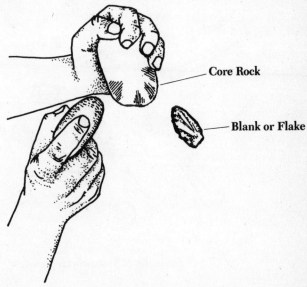

Core Rock

Blank or Flake

The type of rock and the angle of the blow will determine what happens to the rock. If you hit a glassy or semi-glassy rock straight on, it often shatters, leaving a little nodule and a scar of concentric rings that looks like a seashell. This is called a conchoidal fracture—a sure sign that you've got good rock for a sharp blade.

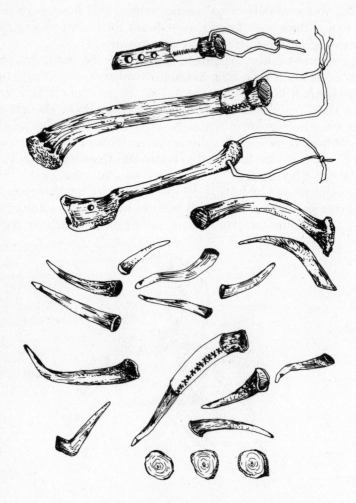

Examples of Stoneworking Tools

Direct percussion is usually done with a rock called a hammerstone. Hammerstones come in different shapes and sizes, from big, softball-sized bashers on down to Ping-Pong-ball–sized pebbles. They can be held in one hand or attached to handles to make hammers. The large ones are used to fracture big core rocks into usable blanks, and the smaller ones are used to rough out the general shapes of tools. You can also use a striking tool called a baton, made from the thick base, or rosette, of an antler.

It takes a certain amount of trial and error to figure out just how and where a rock is going to cleave, so experiment a little to get a feel for it. Start by bashing a few rocks against each other and see what happens. Don't worry if you can't identify the rock you're working with, as you'll learn by "talking" to them. Sometimes you'll find the hammerstone is actually softer than the stone you're working on. If so, it may crack first or you'll have to find a harder rock to get the other one to break.

Once you have formed a sharp edge, don't strike it directly or you will shatter it. Instead, keep striking the rock obliquely on the high points with downward, rolling motions of the wrist. You can vary the angle of the blows, depending on the rock. Generally, the harder the rock, the more directly you can strike it; the more brittle and glassy, the less direct the blow. Keep the wrist flexible, and regulate the power of the blow with your upper arm.

To produce a sharp edge, use oblique blows.

Also, if you're working with glassy rock like obsidian, never work it with your bare hands. The splinters and dust fragments are so sharp that you'll pick up cuts without even knowing where they came from. Instead, wrap the rock in some kind of protective material such as rawhide. If you're working on your lap or leg, put something beneath the rock first. If you're working with the rock in your hand, hold it in a leather or rawhide pad that covers your palm.

Indirect Percussion. When you hit a rock with another one, you're basically sending shock waves through it that break up its molecular structure. For fragile rocks or finer work, you can direct the force of the waves better through indirect percussion—that is, by pounding on a blunt-pointed instrument such as an antler held against the rock at the angle you want the flake to be thrown. First secure the rock beneath your knee or in a crevice or clamp so you can use both your hands. (You can also have someone else hold it.) Then aim the punch and pound it until the rock cleaves. Keep doing this, chipping away at the edges of the core rock until you have some usable blanks.

Indirect Percussion

Pecking and Crumbling. Once you have a good blank, it's a good idea to shape it before knapping. The first step here is pecking—in this case, either holding the blank in your hand or laying it on your knee and striking around the edges with a small rock until it takes on the shape you want. Then you start crumbling the edges to work them down. Crumbling is done by pushing a blunt tool at right angles to the blade.

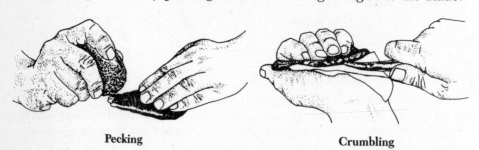

Pecking   **Crumbling**

For a heavy blank you may have to exert a lot of pressure. If you have a thin one, drag the tool more carefully from top to bottom. Do this all the way around the edge.

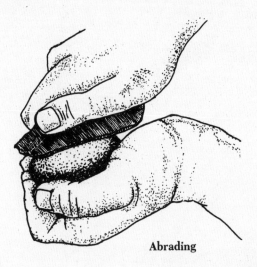

Abrading

Abrading. After the blank has taken on the general form you want, but before the actual knapping begins, rub the sharp edges against a sanding stone. This dulls and flattens the edges and makes a good working platform, or "anvil," for the flaking tool.

Pressure Flaking. Now comes the precision flaking or knapping. This is a very delicate art and takes much care and practice. It also takes a relaxed, patient attitude. Remember, many rock edges are as sharp as razors and can take fingers off as quickly as meat cleavers if you are rushed or anxious and not paying attention.

Pressure flaking is done by pressing a hard, pointed instrument against the flat "anvil" of the prepared blade until the pressure becomes so great that the rock splits and throws off a flake. Many different tools can do this, including rock, bone, wood, and copper or brass nails pounded into wooden dowels. But I find that the best tools are about the same hardness as deer antlers.

The general procedure is as follows: Holding the blank on a protective pad between your fingers and the heel of your hand, place the end of the flaking tool either on a peak or a flat "anvil" area on the prepared edge. Then, with a slight twist of the wrist, push inward so that almost all the force is directed toward the center of the blank.

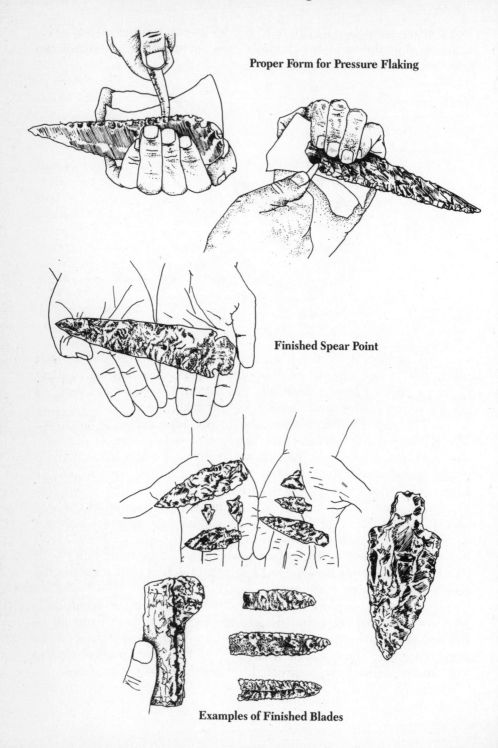

Proper Form for Pressure Flaking

Finished Spear Point

Examples of Finished Blades

This is a difficult movement, but it's critical to the final outcome. When you push against the flaker you're stressing and breaking the bonds that hold the rock together, and the shock wave wants to travel in the direction of the applied pressure. If you push the tool straight down, you'll chip short, little pieces off the edge. If you push inward, the flakes will be long and thin, creating a sharper blade and a more beautifully honed tool. The slight twisting motion also helps to "unzip" the rock to set free a new flake.

Start with a large tool such as a deer tine, beginning at the base of the blade and working toward the end. Keep working along the edge, placing the rounded point of the flaker on the peaks. Throw the first row of flakes as far across the width of the blade as possible. Each time you knock off a flake, you knock off one peak and create two smaller ones, leaving a little scalloped bevel on the underside of the blank. These new peaks in turn can be flaked with smaller tools when you do the next row.

When you're done with one edge, flip the blank over and do the other side. Then pick up a medium-sized flaking tool and work the high spots a second time. I usually have about three different sizes of flakers for finer and finer work. The final sharpening of the blade is done with the smallest tool.

Every rock is different, so you have to get a feel for each one. The core work and the shaping of the blank will help with this, but so will your senses and your intuition. As you manipulate the rock, you'll get a better idea for where the fracture lines are and what kind of pressure to apply. The edge you create with one blow will show you where to take the next one.

One of the reasons that I love working with rock is that it is a communion with an entity that we usually think of as nonliving. But in a way a rock is no more dead than a deer or a butterfly. It does not move or communicate in a way that is perceptible to our senses, but we know through quantum physics that a rock is a web of pulsating energy. At the deepest levels it is only distinguishable from a plant or animal by the density and movement of its atoms. Maybe we do not hear the rock talking because it is not alive. On the other hand, maybe we do not hear it because we do not listen. If you can, treat the rock as a living spirit and see what a difference this makes in your perceptions and craftsmanship.

Grooving. Once the blade is fairly well honed but not too sharp, you may want to groove or notch the tool so that it will fit onto a handle or arrowhead. Do this by crumbling the base of the blade with a punch, and sand the notch before lashing it to the handle.

Boneworking

Almost as great a variety of tools can be made with bone as with stone. In this chapter you've already read about how bones and antlers can help to shape rock into sharp blades, and generally these bone implements require very little work. Bone is not as hard as stone, but it has the distinct advantage of being more easily worked. From bone and antler you can very quickly fashion awls, needles, punches, handles, scrapers, hoes, rakes, clubs, arrow straighteners, and a variety of other tools. Bone even makes serviceable knives, drills, spear points, and arrowheads. All it takes is a knowledge of how to split and abrade the material, plus a little intuition and imagination.

There are several different approaches to boneworking. The skills you use will depend mainly on the type of bone and your own needs. Start by looking at the bone, holding it in your hands, and getting a feel for what it has already been. See if you can visualize what part of the animal's body it came from and imagine it working with the bundles of sinew and muscle that were attached to it.

Then see it as part of the complete animal. Maybe you'll visualize a deer running free in a field. If so, watch how the bone meets the cartilage, absorbing the impact as the deer's hooves dig into the soil. Watch the muscles contract as the deer prepares to spring again. See the bone sailing lightly through the air, free of all tension and pressure as the deer becomes airborne.

Also think about how a particular bone might be used. If you're working with the leg bone of a large animal, you may hold it in your hand and suddenly realize that it would make a perfect knife or tool handle. A scapula bone might suggest the flat, rounded contours of a scraper. Just by their size and shape, most bones will give you a hint about what they want to become and how you can use their natural strength and shape to best advantage.

The support bones of large animals are usually harder and heavier than smaller bones, so they are especially good for handles or for long, pointed instruments—for example, knitting needles or spear points. To make these things, you often have to break the bone into thinner pieces. You can do this simply by scoring the bone lengthwise with a thin piece of sandstone and then tapping it along the cut with a hammerstone, just as you might score and tap on a piece of glass to break it in two.

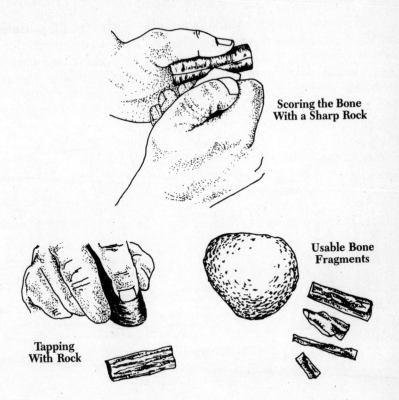

Scoring the Bone
With a Sharp Rock

Usable Bone
Fragments

Tapping
With Rock

Stone abraders and saws can also be used to cut bones crosswise into rings and hollow tubes. From these you can make things like gougers, bird calls, flutes, and jewelry. Another approach is to hit the bone with a hammerstone. This usually breaks and splinters it, leaving a variety of sharp fragments that can be abraded into other tools.

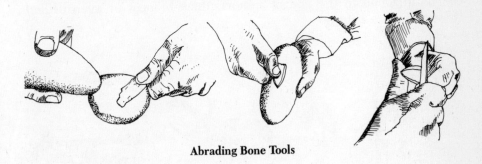

Abrading Bone Tools

Most small bones hardly have to be worked at all, outside of a little abrading. One of the best abrading rocks for sharp, pointed instruments is sandstone that has been grooved by pecking (see page 104). This makes it possible to hone awls and needles to the fine points needed for working rawhide and buckskin.

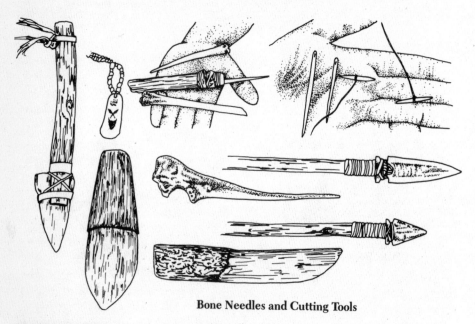

Bone Needles and Cutting Tools

Hafting Tools

Most tools are more effective if they are lashed onto wooden handles. For a stone axhead, you can heat a sapling and bend it around a groove that's been pecked into the rock. For a knife, you might cut a deep notch in a wooden handle and fit the blade inside it. An adze or ax can be attached to the end of a short, thick branch, or even drilled through before hafting.

Hafting an Axehead Onto a Split Sapling

Rawhide. Whatever kind of handle you choose, you're probably going to need cordage to lash it on tight (see also *Cordage*, page 88). I find the best lashing material for large tools is rawhide. It is strong and shrinks as it dries, which binds it more tightly than you could ever do with your hands. To prepare it, first slice it into thin strips and soak it in water until it's pliable and translucent.

If you've got a stone you want to lash between two halves of a handle, place the stonehead between the two pieces. Then, starting a couple of inches below the stone, clamp the rawhide between the two pieces and start wrapping it as tightly as you can without breaking it. Be careful not to wrap around sharp edges. Work your way up to the blade, then back down, then up once more, overlapping the rows of rawhide after crisscrossing over the stone three or four times. Then tie it off and let it dry.

Rawhide Binding. Rawhide is also good for mending and binding. Its strength makes it capable of repairing even broken gunstocks, and it shrinks to conform to the shape of the handle. If you're covering a knife handle or some other tool, cut and shape the rawhide so that it wraps completely around the handle and overlaps itself just slightly. Then with an awl, punch holes in the edges and stitch it up with cordage.

The kind of stitch you use isn't very important, as long as it's tight. A circular stitch with a knot at the end is the easiest, though you can produce a more durable bond with the baseball stitch (a series of overlapping X's). Once the rawhide dries, it almost fuses or melts together at the seams, leaving an extremely tight and beautiful surface.

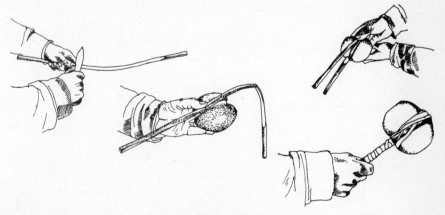

Hafting With a Bent Sapling

7
BOW AND ARROW

In *Tom Brown's Field Guide to Wilderness Survival*, I described many methods of finding and catching animals for food. Among those techniques were traps, spears, throwing sticks, bolas, atlatls, and crude bows and arrows. All these methods are equally effective for long-term survival. However, there is no hunting method that so typefies the earth connection as the finely crafted bow and arrow.

The bow and arrow has a long and beautiful tradition spanning thousands of years. It has played an important part in the survival and development of almost every culture on earth. In many ways it reached the height of refinement among the native peoples of North and South America.

To these peoples, the bow and arrow was not just a means of killing game for food. It was a sacred tool to be made and used in a way that showed respect for all living things. The bow and arrow was always made with care and devotion, the craftsman's prayers going out to each article that helped to produce it. It was a means of realizing the physical and spiritual connection between the hunter and the hunted. The bow was like the great arm of the Creator, and the arrow was like the lightning bolt that freed the animal's spirit and joined its body to the hunter and his tribe.

Times have changed since the days of the sacred hunt. Today, most people hunt for sport, and many of them with little regard for the life that is taken. Even modern bow hunting, which was originally supposed to be a return to simpler times, often involves such an array of pulleys, cables, and other paraphernalia, that it is almost as automated as rifle hunting. As in other areas of life, technology often reduces the need for skill and feeling and creates such a distance between us and our origins that we no longer feel the beauty and flow of the life we are taking.

To me, there is no greater pleasure than making things for myself, and hunting equipment is no exception. This is also part of the earth connection, and you can get it through making a bow and arrow even if you never go hunting. On the other hand, the bow is not really complete without the hunt. There is something you learn about life through using the bow that cannot be learned in any other way. The true essence of the hunt is found in the hunter's relation to the living animal. That is why I am an advocate of tracking, stalking, and close-range hunting.

Many people today think that the Indian bows of old were ineffective hunting tools. This is untrue. When combined with stalking and camouflage, they were as deadly as high-powered rifles. In fact, they were even more deadly because the Indians used them so flawlessly. At close range they were so powerful that it was not uncommon for a hunter to send his arrow clean through the body of a giant woodland bison. Until the repeating rifle became common, many American mountain men found the bow and arrow more reliable than their manufactured firearms.

The bows and arrows described in this chapter, then, are not toys or museum pieces. They are serious and highly refined survival tools. If they are made and used with skill, they will kill as effectively and humanely as any modern bow and arrow on the market.

But mere survival is not the reason for this chapter. I am not interested in helping people kill more animals. I want to generate a deeper appreciation for them. I want people to feel the very real sacrifices that plants and animals make for our livelihood, and to realize our beautiful connection with them.

Bows in General

First, a quick look at the bow and its different parts. Each bow is made up of two "limbs," or halves, which meet in the middle at the handhold, or grip. The part of the bow facing away from the hunter is called the back. The part facing toward the hunter is called the belly. Many bows are reinforced with strands of sinew laid along the back. This is called sinew backing.

The ends of the limbs are called ears. With a straight bow, the limbs are nearly straight. When such a bow is unstrung it looks like a long, smooth stave. With a "curved" bow, the ears of the unstrung bow are curved toward the back. In a "reflexed" bow, the entire length of the limbs is curved toward the back. A bow that is both reflexed and curved looks like a boomerang with curled ends. Curving and reflexing are used especially with small bows to add greater strength, snap, and cast.

The strength, or draw weight, of modern bows is measured in foot-pounds of pressure when the bowstring is drawn back to a distance of twenty-eight inches from the belly. However, the effective strength of the bow is a combination of "snap" and "cast." The snap is the speed with which the arrow leaves the bow when you release the string, while the cast is the horizontal distance it can travel through the air.

Over the years I have experimented with many different kinds of bows, and through trial and error I have come up with a hybrid of native American types that I find works with the greatest variety of hunting techniques, terrains, and weather conditions.

Since I am a close-range hunter, I prefer a bow that is curved, or eared, like some of the Eastern Woodlands bows, but somewhat smaller and with sinew bowstring and backing for greater snap and cast. A small bow is excellent for stalking through heavy brush and getting close, tight shots. I have adapted this feature from the bows of the Plains Indians. Plains bows were very short and powerful—both curved and reflexed for greater strength while hunting buffalo from horseback. In rainy weather, on the other hand, I am forced to use a somewhat longer, curved bow without sinew and with a plant fiber bowstring to resist the dampness. When I go bow fishing, I prefer a long straight bow.

I will begin with an explanation of how to make a simple but finely tuned straight bow. Then I will describe how to curve the ends to increase the snap and cast. Finally I will explain how to put sinew backing on the bow to give it greater strength and longevity. This should give you the basics of fine bowmaking. Then, through a combination of experimentation, field work, and museum research, you'll be able to use these same methods to create a bow that truly fits you and your own hunting technique.

As with most tasks, experience is the best teacher. When asked how to do something, most Indian teachers (especially the older ones) used to say, "You figure it out." This is good advice. You end up making some mistakes, but you don't know what's right if you don't know what's wrong. So jump right in, make a few mistakes, and keep after the art until you have mastered it.

The making of a fine bow is a long-term project. There is hardly any way you can do a good job if you are concerned about time. If you get impatient, remember that some of the finest native American bows took several years to make. A good friend of mine who is the bowyer for the Seneca tribe often spends from three to four years to construct a fine bow, and I have heard that some Turkish bowyers used to spend as long as ten years on their bows from start to finish.

When I make my bows, I never rely on tapes and calipers for measurements. Like native peoples, I prefer to use simple hand and eye measurements. Everything I do is by touch, sight, and inner feeling.

Therefore, the dimensions given in this article are by no means law. They are just general guidelines to help you in making your first bow. After you have made a bow or two, you will also begin to rely more on sight and feeling for measurements, thus producing a bow that is as personal as your picture. As a matter of fact, once you become well attuned, you'll discover that the entire process can be done just as well this way as with tape measures and other tools. When this happens, it's a sure sign that you're learning to "read" the wood in much the same way earth peoples do all over the world.

The bows and arrows described here can be fashioned with tools made from natural materials. But let's face it, we're modern people. We have different tools. If they work for us, there is no reason we shouldn't use them. Later you may want to become a purist. However, for ease in learning, I would recommend starting out with a few simple but modern tools. These include a drawknife, a large "bastard" file, a spokeshave, a planer, a cabinetmaker's shave, a mill file, a knife, and various grades of sandpaper.

Choosing the Wood

Generally, the different bowmaking woods used by the Indians were native to the geographic regions in which they lived. However, it was not uncommon to hear of hunters traveling hundreds of miles to get the wood of their preference. The Indians considered these trips well worth the effort, because a good bow could easily last a lifetime and was often passed down from generation to generation.

The native woods for bowmaking are Osage orange, yew, ash, and hickory. Osage orange and yew are especially good because their fibers are long, elastic, and tightly packed together. In some areas of the country, other wood types were also used. These include honey locust, mountain mahogany, greasewood, ironwood, and even cedar and juniper. If one of these trees does not grow in your area, you should be able to get it through a specialty hardwood dealer. Be careful to select only fine, straight-grained wood.

The bow is usually made from a sapling or tree branch. When you harvest the wood, look for a stave that is about a foot longer than you want your bow, to allow for checking and cracking during the drying process. A good general length for a long straightbow sapling is about five feet; for the short bow, about four feet. But you can make them any length from three-and-a-half feet to over six feet, depending on your preference and hunting style.

If you are cutting a sapling, look for one that is two-and-a-quarter to three-and-a-quarter inches in diameter, free of knots and blemishes, and very straight. I prefer to cut my saplings in February when the sap is down. It is more difficult to find a straight, strong tree branch, but it's good if you can, because then it's not necessary to take the life of the whole tree.

Wherever you harvest your bowstave, remember that the wood is a gift from the Creator and should be taken with respect. I never go out to harvest a bowstave without giving it a great deal of forethought. I always look for a crowded grove of young trees. The lack of sunlight forces them to grow straight and tall, with closely packed growth rings. This makes the bow stronger and springier. In other words, I want a sapling that has been under pressure, and ideally one that isn't going to make it. In this way I can get the bow I want, I can prevent the waste of a good sapling, and I can even improve the growing conditions of the saplings around it.

But there is more to taking a bowstave than meets the eye or the advantage of the hunter. There is also the respect. I never go up to a bowstave sapling without looking, touching, smelling and talking to it. In a wordless communication, I open my heart to that which I am about to remove from the flow of life. If I could translate, it might sound like this:

"Wood, your roots reach deeper into the earth than my own. For many years you have watched over this ground. You have saluted the sun as the seasons have circled. You have seen plants and animals come and go. Your branches have nourished birds and insects. Your leaves have gathered and passed on the light from the Creator. Now I ask you to make the ultimate sacrifice, to give your life so that I may live more fully.

"Tree, I choose you to become my bow. Your struggle has made you strong, but you cannot win your battle here. Let me take you and put your power to a new purpose. You who have so long fought for life will now be used as a seeker of death. But be not disturbed, friend, for the death you will seek is but a change of worlds. The end is only new life. Just as your branches now reach skyward, so will your spirit soon follow in their path. Where now you bend and sway in the wind, soon you will bend and sway in my hands to send your arrows toward the sun."

Of course I do not say all these things when I cut a sapling. The words are not important; it is the *feeling* that counts. It is a communion with a living entity, and the benefit goes both to the tree and to the person who is taking it.

After I have cut the sapling, I may continue in a similar vein: "Thank you, friend, for giving your life. If you were to die in this woods in a natural way, your body would decay and go back to the earth in its own time. You would nurture the woodpeckers and the boring insects and finally fall to earth and decompose into the bosom of Creation. I have now removed you from that path. Now, instead of bringing new life to this forest, you will be transplanted to my world and bring new life to the forest of my people."

If I have bought the wood from a hardwood supplier, it makes no difference. I still run my fingers over the wood, thinking about where it might have grown and what it might have experienced. Sometimes I imagine I can see exactly where it grew. I can feel its spirit, and that spirit becomes a part of me. This is what is meant by putting your own personal "medicine" into an implement. You are not just making a bow; you are investing it with an intimate part of yourself. Anything made in such a manner means so much more to the maker than it ever could if it were simply bought in a store or hacked down without any forethought. If treated with the proper attitude, it becomes a part of your living essence.

Even from a purely utilitarian viewpoint, this kind of communication makes good sense, because it helps you take more care in the construction process. The attitude is not so different from that of the skilled artisan or sculptor who simply loves the smell and texture of the wood he works with, who sits and spends hours examining it, trying to discover the hidden forms within it and hoping that his skill will be up to releasing the spirit of those forms.

Seasoning. Once you've cut your stave, it should be seasoned for a year. That's right—a year. If you're lucky enough to get a lightning- or fire-killed tree that's already seasoned, or wood that's been sitting in a covered lumberyard for a year, fine. But with a freshly cut sapling I usually put the stave in a woodshed or other cool, dry place for three to six months, then bring it inside for another six.

The stave should be kept in a cool, dry place even during the inside seasoning, and it should be hung vertically from one end so that it won't warp or bend. You might want to suspend your bowstave from a garage ceiling with a piece of cordage. A good thing to do inside an earth shelter would be to suspend it just behind the smoke hatch. Some people even recommend covering the ends of the stave to prevent them from checking. But I find that if the wood is going to check, hardly anything can stop it. So I leave the ends open, shaving off about an inch of the bark on either tip.

Once the stave is seasoned, remove all the bark. Do not carve it away; scrape it off, holding a knife or scraper at a ninety-degree angle to the wood. This will prevent slipping and cutting into the grain. At this point I usually let the stave season another two weeks before splitting it, just to make sure all the moisture is out.

Splitting the Stave. If you're using a small sapling and you can split it down the middle along its full length, you may be able to get two bowstaves from a single piece of wood. A safer method is to shave the sapling with smooth, even strokes from a drawknife. When finished, bows made from small saplings are semicircular in cross section.

If you are using a larger sapling or small tree, split it in half with a

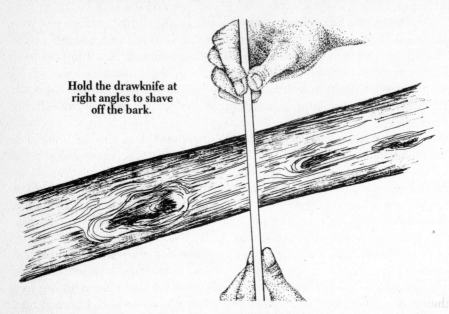

Hold the drawknife at right angles to shave off the bark.

knife or wedge and use the same shaving process. You may be able to get as many as four bowstaves from a single four-inch tree, but I don't trust my second split and usually finish the job with a drawknife.

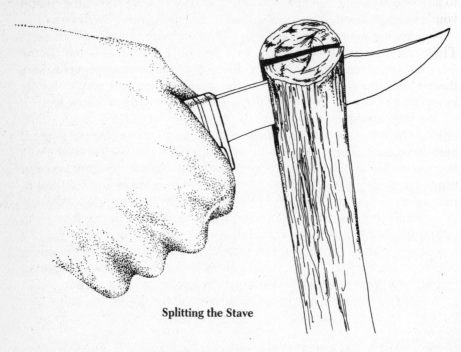

Splitting the Stave

After you have shaped the stave with the drawknife, never carve on it again. I can hardly emphasize this enough. Carving tends to cut into the grain, which may make the bow prone to splitting later on. Further shaping should be done only with filing and abrading tools.

Now you are ready to start work on the bow. Before you do, it's a good idea to ask the bowstave what it wants to become, rather than making it what you think it should be. Consider the grain, the quality of the wood, and the general growth patterns to help you decide how to shape the bow.

The grain especially will speak to you. Stand back from the stave and imagine where the bow lies inside it. You want to cut it so you will get the most power out of it. The strongest grain should be on the back of the bow and the weakest grain on the belly. This is easy to determine. The tighter growth rings are always in the sapwood, in the outer part of the tree, while the looser grain is in the inner heartwood. That is, the back of the bow should face outward and the belly inward.

Measuring the Stave. Generally you should start with a stave that is about two-and-a-half inches wide and one inch thick. If you're using a softer wood, it's best to make a wide, thin bow because it will better resist cracking. Hardwood bows can be a little narrower and thicker from belly to back because they aren't so brittle. If you're going to put sinew backing on the bow, it makes little difference which shape you choose, since the backing will reinforce even most weak bows.

The Self Bow

The self bow is the simplest kind of fine bow you can make, and the techniques used in making it form the foundation for every other kind of quality bow, no matter what the shape or finished design.

Measuring the Stave. You can measure the length of the stave either with a tape or by feel. However, I find that bows fit an individual better if they are measured with the arms. When I measure my bow, I bend over the ground or worktable with the middle of my chest over the grip area, and with my arms and fingers outstretched toward the ends of the stave. If I want a long bow, I mark the stave at the ends of my fingers. If I want a short bow, I mark it at the wrists.

There are other ways of measuring, depending on how long a bow you want. The Plains Indians, who had very short bows for hunting buffalo from horseback, sometimes measured from the right shoulder to the tip of the fingers of the left hand, or simply from the waist to the ground. This made the bow anywhere from thirty-five to forty-eight inches long. You may discover some other method works best for you.

Once I have measured and marked the ends of the bow, I find its center point. From this line I measure out about three inches in either direction to mark off the grip area. Another way of making this measurement is to grasp the bow with your hand over the center and mark the ends of the handle about a finger's width on either side of your hand.

Holding the Stave. There are many ways to hold the stave as you work. One of the simplest is to use a small, adjustable vise table. Another is to use the power and leverage of your own limbs and various natural objects. For example, you can create a quite effective grip by sitting in a cross-legged position and locking one end of the stave under one knee and laying the other end over the opposite thigh. You can also work the limbs fairly well from a standing or kneeling position by bracing one end of the bow against a padded rock with the other end against some kind of padding on your chest. Other approaches also afford good leverage—for example, leaning the stave against a tree trunk and straddling it as you work.

You can also make a number of simple vises from natural materials. One of the simplest is to sandwich the stave between the flat, notched surface of a tree stump and a block of wood that is held down by a rock. Leverage can be applied from the side of the stump by a long stake that is beveled to fit against the edge of the block of wood. A similar setup can be made by notching a log and wedging the long stake between the wood block and a tree trunk. You can make a set of small vise clamps with two blocks of wood lashed over a dowel, too, and force them shut by driving a wooden wedge in at the other end. You can even make a log vise by placing one heavy log on top of another and weighting it down with a rock. Ingenuity can make up for lack of tools, whether you're in a survival situation or not.

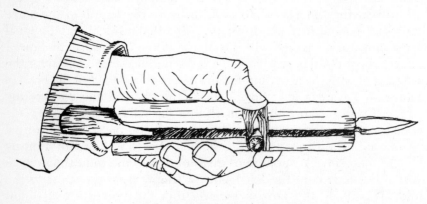

Hand Vise Holding Bone Arrowhead

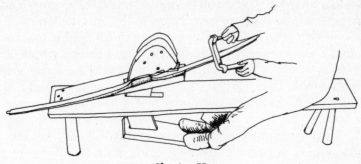

Shaving Horse

Finally, one of the most useful bow-holding devices is the shaving horse, a kind of workbench with an adjustable jaw and foot pedal. You can make one of these contraptions quite easily. With it you can clamp down or let up on the stave anytime you want. This allows you to shave and file and move the stave around to any position without damaging the bow.

Survival Workbench for Bowmaking and Woodworking

Rough Shaping. Once you have marked off the grip area and you know the general shape of the bow, begin to thin out the limbs. Starting from the outside of the grip lines and using a rough honing tool such as a bastard file, Sure Form, or sanding rock, begin thinning and tapering the bow from about a five-eighths-inch thickness near the grip to a three-eighths-inch thickness at the tips. The width should also be evenly tapered, from about two-and-a-half inches near the handle to one-half inch at the tips. Don't get carried away and try to taper the limbs like a store-bought fiberglass bow, as this will only weaken them. What you want is a smooth, even tapering from handle to tips. The thicker and wider the bow, the stronger the pull.

When the tapers are all relatively smooth, finish the abrading with a finer tool such as a mill file. As you work, try to get a feel for the wood, using the tool with a rolling motion. After a while you will know just which tool to use, just as you instinctively know which teeth to use when chewing.

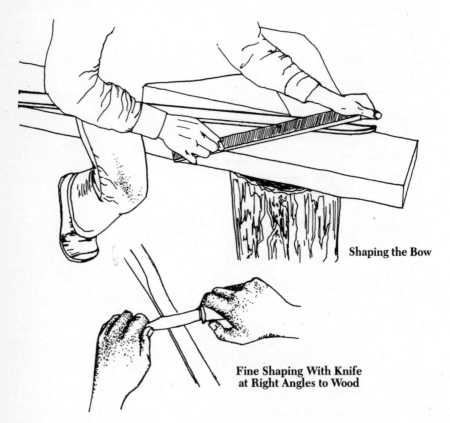

Shaping the Bow

**Fine Shaping With Knife
at Right Angles to Wood**

Be sure to eye the bow from all angles. Look from the center to both ends, both ends to the center, from above and below, and from other angles. Check for high and low spots. At any time you should be aware of whether the bow is being formed symmetrically and whether the tool is flowing with the grain of the wood. Also be calm and methodical, never impatient or hurried. Your own moods and feelings will become a part of your bow, and you want it to reflect a positive, loving attitude. This is the hallmark of the skilled craftsperson.

When you're done with the limbs, begin to work the grip. The grip is mainly a matter of personal preference. It should be shaved in width and thickness until it fits your hand and allows room to place the arrow. As you work the grip, grasp it from time to time to see how it feels. Once it feels comfortable and all the lines and tapers are relatively smooth, finish the abrading with a finer tool such as a mill file.

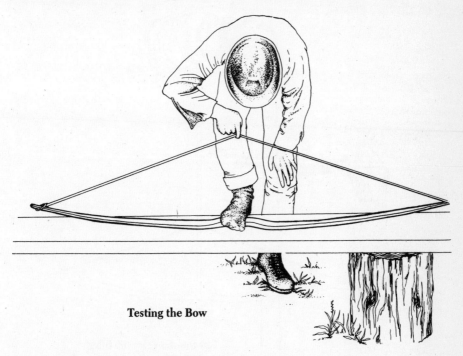

Testing the Bow

Testing and Fine Tuning. The bow must be tested to see if the limbs pull evenly. The simplest way of doing this is by tying a string to the ends, placing your foot in the middle of the bow grip, and pulling slightly upwards on the string. As you pull, have someone looking to see whether the limbs bend evenly. Be careful not to pull the ends up more than an inch or so at a time, as this may cause splitting if the limbs are uneven. If one limb pulls more easily than the other, carefully abrade the belly of the stronger limb, using a small mill file. Keep testing and abrading until both limbs pull evenly.

Tillering. A safer and more accurate way of fine tuning is to use a tiller tester. The tiller is a piece of wood about three feet long that has a row of evenly spaced notches every two inches or so for a distance of twenty-eight inches and is shaped at the lower end to fit against the belly of the grip. (You can also carve a stationary tiller into a log or pole, with a deep "bow-hold" carved just below the first notch to take the belly of the grip. The vertical tiller makes it especially easy for a lone person to gauge the bend of the limbs, because you can leave the bow in the tiller and go around to the side to eyeball it.)

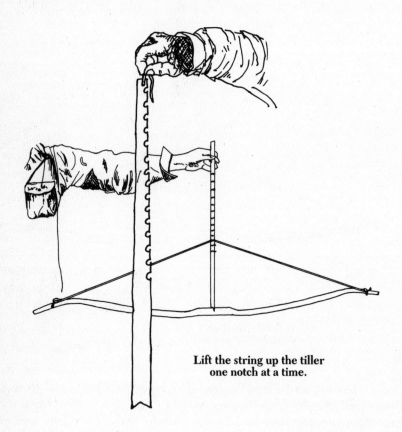

**Lift the string up the tiller
one notch at a time.**

Once the grip is in place, the string is pulled up to the first notch and the limbs are observed for evenness of pull. If they are even, the string is brought up to the second notch, and so on, until the bow has been tested and honed all the way back to the twenty-eight-inch mark.

Shave the belly very gently on both sides of any weak spots.

During this process, if you spot any irregularities in the pull, shave carefully on both sides of the weak spot (on the belly *only*) to even it up. Then shave the opposite limb so it matches the strength of the other. Do this at each notch—and do it safely—by taking the bow off the tiller beforehand. This will protect you if the bow decides to snap. Also remember that the finest shavings make a big difference at this stage. Do not use a rough honing tool, but only the finest blade or even a piece of sandpaper.

If the bow is inherently weak, it's possible that it will snap during this process. If so, you will just have to steel yourself against failure and try again. If it passes this test, I would suggest testing it for strength with an inexpensive bow scale. Most state regulations require a minimum thirty-five-pound pull (or draw weight) on hunting bows to make sure they are powerful enough to kill an animal. If you have a small bow and your draw length is less than twenty-eight inches, you will have to allow for that.

You will also have to allow for changes in strength brought on by changes in the weather. When it's damp and warm out, the draw weight decreases because heat and moisture absorption make the wood more pliable. When it's cold and dry, the bow becomes less flexible and will have a higher draw weight. Generally, you should allow for a variance of ten pounds on either side of the required weight. Though this sounds complicated, it becomes almost a matter of feel after you've made a couple of bows.

Once you have tuned the bow, sand both the limbs smooth. To keep the bow supple and semi-waterproof, it should be oiled with fat or brains. The native Americans used bear or deer fat that had been rendered and applied it warm. Deer brains were also used after being warmed up in much the same way as the tanning leather (see *Hide and Hair*, page 155). The bow was then set close to a fire to warm the fat and drive it deeper into the wood. My personal preference is to mix rendered fat and deer brains and apply the mixture warm, then set the bow high above my fire or wood stove to drive in the oils.

When this is done, the bow is ready to shoot. Go easy at first. Sometimes, no matter what you do, the bow will snap. This may be through no fault of yours, but because of inconsistencies or flaws in the wood. If this happens, there is nothing to do but try again. After all the work that goes into the bow, it can be heartbreaking to have it snap on you during the first few tries. This is why I recommend reinforcing some bows with sinew backing (see page 135).

Curved and Reflexed Bows

To add more snap and cast to your bow and to make its performance more volatile, you may want to curve (bend toward the back) the last six or seven inches of the limbs. This should be done just before you do the final tillering. Using a pot of boiling water, immerse one end of the bow up to about the nine-inch mark, and let it sit for two-and-a-half to four hours. (At home, I use my wood stove and add humidity to the house at the same time the bow is cooking.)

While the bow end is in the water, cut out a curved pattern on a two-by-four-inch piece of scrap lumber and make a set of wooden forms, as shown. The ears of the bow will be sandwiched between these and clamped down to cool and dry. The degree of the bend depends on personal preference. Anywhere between forty-five and ninety degrees is fine.

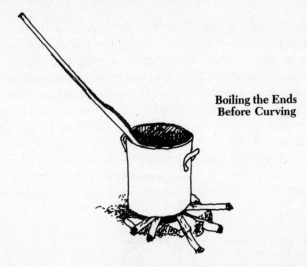

**Boiling the Ends
Before Curving**

Once the bow end is removed from the kettle, place it onto the form and clamp it down. The best way to do this is to use two clamps, or vises, and clamp the tip of the bow first. Then, using the bow itself as a lever, slowly bend the rest of the ear back over the form and put on the second clamp. It's a good idea to place some kind of padding on the

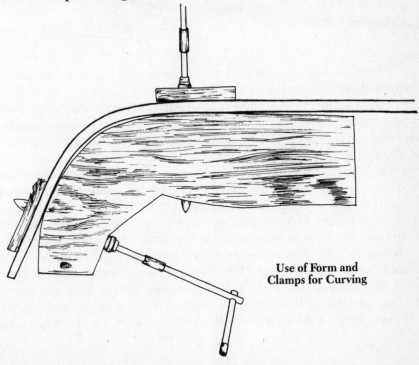

**Use of Form and
Clamps for Curving**

inside of the forms to keep the bow from getting dented. When you have one ear clamped down, boil the other and repeat the whole process. Then let the bow dry for a day in a cold, dry place.

There are several other methods you can use to bend the ears. One is to steam and bend the wood in stages rather than putting the clamps on full force. In other words, steam the ear and bend it slightly, letting it set and cool between the loosely clamped forms. Then steam it a second time and tighten the clamps a little more. Finally, steam it a third time and tighten the clamps all the way. This method will be less likely to split the fibers. If you're out in the woods, you can also make an effective bow clamp by tying the steamed ends around a tree.

**Steaming Method
Using Primitive Kiln**

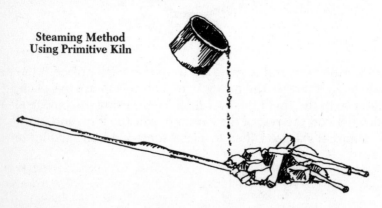

Another bending method that's especially useful in wilderness situations is to soak the ears in water for a few hours and then steam them inside a tiny sweat lodge supplied with hot rocks. The sweat lodge is just a few feet in diameter. The end of the bow can be pushed in one side, and water can be poured on the glowing rocks from a small opening in the top.

If you don't want to go to all this trouble, you can hold the soaked wood over hot coals until it's hot to the touch. Or make a primitive kiln by hammering a heavy pole into the ground at such an angle that it goes beneath your fire. Pull the pole out and mound up some dirt between the fire and the hole so you don't burn the bow. Then dump water down the hole and leave the bow there for a few hours with the fire going. When the ear is pliable, bend it to the desired shape and clamp it down.

Sometimes the back of the bow will split a little during the bending process. If this happens, don't despair. A minor split may be only superficial; it does *not* mean the bow is ruined. On the other hand, there are some bows that are going to split and crack no matter what you do. It's just one of those things you have to be prepared for.

Fine Tuning. The next step is to remove the clamps and fine-tune the ears. This is done by removing a little of the belly wood just in front of the ears. I usually take off one-sixteenth to one-eighth inch of wood from the start of the curve to a point about six inches back. I find that by doing this the bow has better snap with very little jolt. Some bowyers say that it also keeps the ears from snapping off. When you are done with this, proceed to fine tune the limbs as explained on page 129.

Once your bow is fine tuned, you can either sinew-back it or finish it the same way you did the straight bow. If you want to go further, you can add a reflex to your bow by bending the back at the center. Just heat the center of the bow over a steaming kettle for two hours or more. Then lay the back over a small log and stand on the ends or weight them down until the bow cools. The backwards curve this produces will add even more power to your bow—especially if it is a short one. Anytime you reflex and recurve a bow, always reinforce the back with sinew. If you don't, shooting it may put such a strain on the wood that the bow will snap.

Sinew Backing

Regardless of the length or style of your bow, backing it with sinew will increase its performance and longevity. It will also add an automatic reflex to it, because the sinew shrinks and pulls as it dries, creating a backwards arc in the limbs.

The two necessary ingredients for this process are sinew (fiber bundles from the leg or back tendons of an animal) and hide glue (made primarily from hide and hoof shavings). I prefer the leg and back sinew of deer or elk, but I have found that horse, goat, and moose sinew work almost as well. Generally the hoofed animals will provide the longest and thickest bundles of sinew, and invariably the sinew from a wild animal is stronger than that from a domesticated animal.

To obtain the sinew, simply remove the tendon bundles from the back and legs of the animal. The legs of most hoofed animals contain very little meat. The leg bones are basically round in front, but in the back there is a shallow, U-shaped trough. The sinew rests right in that trough. Back sinew lies in long, white cords that run along both sides of the backbone. It's usually found just beneath the flesh and can be lifted out very easily. The backstraps are excellent for bow backing and bowstrings (and also for sewing) because they are composed of such long

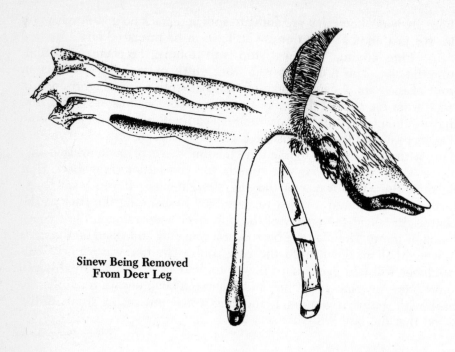

**Sinew Being Removed
From Deer Leg**

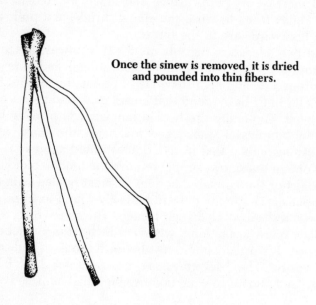

**Once the sinew is removed, it is dried
and pounded into thin fibers.**

and elegant fibers. The leg fibers are coarser, more bunched together, and not as graceful as the backstraps, but they are somewhat tougher and more powerful because they've had to support the animal's weight and propel it through its daily activities. For this reason they are harder to break down, but they also give more strength and are excellent for the final sinew coatings on the back of the bow.

If you aren't a hunter, contact your local animal rendering service or slaughterhouse and arrange to buy some sinew. Usually they will give it to you for nothing, though they might also give you a few strange looks. There isn't much public demand for animal sinew, and most people wouldn't understand even if you tried to explain it to them.

Nor will most people understand that you can feel a very real connection to that animal through the sinew. Yet there is hardly a time I pick up a bundle of it that I don't in some way feel the life flow of the animal it once belonged to. First there is a physical beauty to it. I like to feel it in my fingers, sensing the thousands of fibers within it. I like to feel its tensile strength and watch the light shine through it. I also like to imagine how all those fibers might have worked together to send the deer springing over a fallen log or dashing across an open field. The colors, textures, and even the smell and taste of the sinew becomes important to me when I'm working with it.

Why do I take the trouble to visualize these things? Because I am going to use this sinew to strengthen my bow, in much the same way that it once strengthened the backbone of the deer. Or perhaps I will use it to sew the clothes that will keep me warm, or to bind feathers and stone heads to the arrows that will supply my food. I am going to transplant this sinew into the very fiber of my life, and there is something in me that needs to welcome it and to know and express thanks to the owner who has given me this gift. I feel this way regardless of what part of the animal is used. I am welcoming the presence of the animal into my life. I am also utilizing the animal in a sacred manner so that I can put things into perspective and not be wasteful of what has been given to me.

Again, this thankfulness is not something that is expressed in words. It is only felt in the heart. But it is a vital part of the process. With continual prayer, the connection to your tools and materials is made and maintained, whether it comes during the taking, preparation, making, or use of the finished product. And you needn't reserve this communication just for "natural" materials. In the end, everything is natural and connected to the earth.

With this frame of mind, prepare the sinew by scraping off the clear sheath that holds the bundles together. Then flatten the bundle out and set it high over a stove or fire to dry. If the bundle is thick, you may have to slice it down the middle like a watermelon and lay the two halves out flat to dry. Whatever you do, don't dry the sinew in direct sunlight or high heat, as this may cause fat burning.

When the sinew is dry and hard, pound the bundles with a wooden mallet or soft rock against a thick board or workbench. (Don't pound the sinew between two rocks or you'll fray and break the strands!) This pounding process takes a lot of time and patience, but it gradually breaks the bundles into bands of whitish-gray threads.

**Appearance of Sinew
After Drying and Pounding**

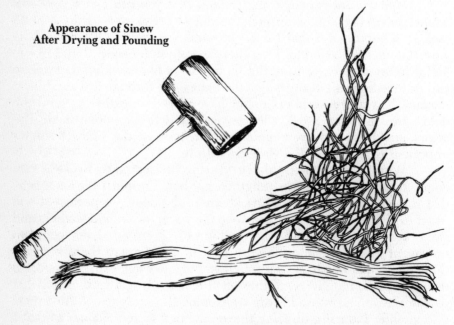

If you're impatient, mull over what you're doing. Listen to the rhythm of the pounding mallet. Feel the action of your arm strengthening your own muscle and sinew as you pour your energy into the bundles on the wooden anvil. Think about how they will later be rewoven with your own hands to give strength and solidity to a tool of your own making.

The rough sinew produced in this way can be shredded and used right away for bow backing or wrapped into cordage for bowstrings. To get a higher grade sinew, you can take a little more trouble to clean off the excess fat after taking it out of the animal. This can be done in several ways. One is by wiping it quickly with a solution of tannic acid. Another is by dipping it into a solution of water and white wood ash and then wringing it between your fingers. You can do the same with a mild soap solution—ideally a mixture of water and meadowsweet, yucca root, or some other plant that contains saponin.

For a higher grade sinew, be more careful to separate the fibers during the pounding process. And, if you want really soft, pliable sinew, I suggest buffing it around a small branch or twig. The heat and movement help to loosen and separate the fibers, giving them a soft, cottony consistency. This takes more time, but it makes the sinew much easier to work with.

Hide Glue. To prepare the hide glue, put bits of rawhide, hooves, dewclaws, hide scrapings, and chopped viscera (white and pink organs only) into a pot of water and boil for several hours. For a finer consistency, skim off the oils and debris that bubble to the top. This oil, by the way, is called neat's-foot oil, and it has a variety of uses, from lubricating bowstrings to lighting lamps.

You may also want to strain the liquid from time to time through dried grasses or cheesecloth. After several hours of boiling, you will end up with a thick, gooey mass about the consistency of molasses. This is the hide glue. The best glue, which is very uniform and requires no straining, is made from hide scrapings (see page 164). If you don't want to go through this process, another alternative is to buy commercial hide glue at your local hobby shop.

Backing the Bow. While you're working on the bow, it's best to keep the glue in a water bath over a stove or hot coals so it stays hot to the touch. I find that the flames from a fire don't produce an even enough heat, and if it's at room temperature it gets gummy and sets up too fast. If you keep the heat even, both the consistency and drying time of the glue are more predictable.

Prepare the back of the bow by scoring or sanding it with a file or a hard, abrasive rock. This gives the glue and sinew mixture plenty of nooks and crannies to adhere to. Then paint the bow back with a light mixture of hide glue and water. If you feel that you may have gotten hand grease on the bow, lightly sponge it off with a detergent-and-water mixture and allow it to dry; then apply the glue-and-water coating.

While the glue is setting, place a little of the sinew in warm hide glue for about five minutes to loosen it up. Then take it out, squeegee it between your fingers, and transfer a few strands at a time onto the bow. (Never let prepared sinew soak for longer than five minutes or it will get too warm and start to curl up.)

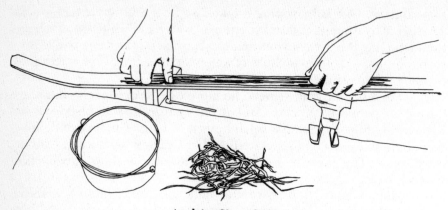

Applying Sinew Strips

Now, starting at the center of the bow and working outward, apply strips of glue-soaked sinew to the back of the bow. Make sure you apply the strips parallel to the grain of the wood, and stagger them to avoid any weak seams. Keep pasting down the strips, smoothing them out with your fingers as you go. Put them on all the way out to the bowtips, then over onto the belly side for about two inches. (Also have a bowl of water handy so you can clean off your gooey fingers from time to time). When you're done, the sinew should look like a mosaic of interlocking bands covering the entire surface of the bow.

Once the bow is covered with the first coat, go back and fill in any gaps. Next, stop and let the first coat dry for a couple of hours until it feels tacky. Then put on two to three more coats of sinew. With successive coats, it's not so important to keep the sinew strands flat. In fact, you can begin to mound them up more, using more leg sinew and somewhat thicker bundles to create a slightly domed effect in the middle. But keep overlapping the strands, always smoothing them out with your fingers. When you've applied all the coats you want, hang the bow up to dry for three to six weeks.

When the sinew is thoroughly dry, you may be interested in applying a final cosmetic coat, such as glue mixed with powdered clay or powdered seashells. Some people even glue a rawhide strip to the belly. These things don't give the bow any more substance, but they do help to prepare the surface for painting.

After the bow is dry, gently file and sand the back until it's smooth. (Sinew can be filed and sanded just like wood.) Then fine-tune it.as explained on page 129. Finally, treat the bow with rendered fat or deer brains in the same way as described for the straight bow (see page 132). This time, however, do not place the bow near the fire for deep penetration of the oils, as it may damage the sinew.

Nocking

There are many different kinds of nocks, or notches, you can use to secure the bowstring. One of the most common is the diamond-shaped nock used by the Seneca, which is filed at about a forty-five-degree downward angle an inch or so from the tips of the bow. Some people prefer to shave off even more of the tips, leaving little protrusions of various shapes with little ledges at their bases for the string to rest on. This is mainly a matter of personal preference. But file the nock gently at a natural angle, and don't cut in so far that you weaken the bow tip.

Protecting the Bow

Sinew is very strong, but it loosens and stretches when it gets wet. This can weaken the action of your bow. Whether or not you use a protective covering, be sure to oil the bow twice a year to keep it supple and in good shape. Some people like to finish their bows by gluing on snakeskins for protection from bad weather. This is a good idea, especially for bows with sinew backing. But once again, it's important to remember that you're taking a life. A snake is no less important and precious to the Great Spirit than any other living thing, and it should be treated with the same reverence and respect.

Bowstrings

The best natural bowstrings are made of sinew, though you can also use many other materials, including rawhide, animal intestines, and various plant fibers. Plant fibers are weaker but better for use in the rain, since rawhide and sinew tend to stretch when wet. Velvetleaf, dogbane, nettle fibers, or the inner bark of the American elm are some of the best plants to use.

Probably the simplest method of making a natural bowstring, whether sinew or plant fibers, is the reverse wrap method (see *Cordage*, page 88). If you're using sinew, chew on the fibers (but don't swallow). Saliva mixes with the natural glues in the sinew to make a very strong adhesive. This binds the fibers to each other and makes a very strong, hard bowstring.

Using the reverse-wrap method, splice and wrap until the cordage is at least twice the length of the unstrung bow. Then double and wrap it again. Some people even do a triple reverse wrap, though the string can begin to get a little bulky.

You can also make strong bowstrings by braiding together several long strands of rawhide or several strands of reverse-wrapped cordage. (Wrapping and braiding can both be made easier by tying one end of each strand to a peg planted firmly in the ground.) To keep the ends from fraying, either weave them back into the twisted cord, secure them with simple overhand knots, or "whip" the ends by wrapping and tying them off with smaller pieces of cordage. This last method is especially useful for bowstrings because you can whip one end and tie it permanently to one end of the bow, then whip the other end after forming a noose that fits right over the bowtip.

To protect the bowstring I like to coat it lightly with a little neat's-foot oil, lard, bear grease, deer grease, or even baby oil. This adds suppleness to the fibers.

Arrows

The Indians knew that a bow was only as good as the arrows that were used with it. Many of us, while looking at museum displays of native American bows and arrows, have come away with the mistaken impression that native arrows were rather poorly made. This was hardly ever the case. The fact is that most of the arrows in museum displays are a hundred or more years old, and they have been warped and twisted as a result of heat, moisture, and gravity. Then, as today, an arrow had to be absolutely straight, and it had to be fitted with feathers and heads that were so carefully honed that they seemed to grow right out of the shaft.

Selection of Shafts. An arrow shaft has to be light, strong, and resilient. Lightness gives the shaft the ability to rocket through the air with little resistance. Strength is important so the arrow can withstand jarring impacts. And resilience gives the shaft the ability to bend but not snap under unusual strain, such as when a stricken animal rolls over on it.

Some of the best woods for arrow shafts are cherry, serviceberry, ash, dogwood, cedar, high bush blueberry, cane, and reed. Cane and reed shafts, being hollow, are the lightest and fastest of all, though they are quite fragile. A wood such as oak would make a very strong shaft, but it would also be quite heavy. Cedar is light and strong, but it tends to snap quite easily, as does willow.

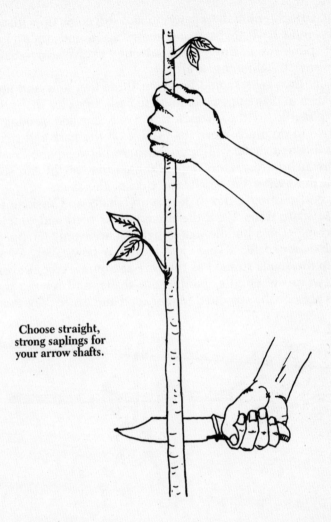

Choose straight,
strong saplings for
your arrow shafts.

Most shafts are selected and gathered in the same way. Choose only the straightest ones you can find. In most cases, you'll search out a thick grove of small saplings or reeds, ideally in the winter when the sap is down. Use the same care and respect when collecting arrow shafts as you did when choosing your bowstaves, always looking for a place where the plants are growing in fierce competition and choosing mainly the ones that aren't going to make it.

Take your time. I can't emphasize this enough. I could easily select twelve arrows in half an hour, but then I might have to spend twelve hours straightening them later on and still come up with shoddy arrows that would have to be straightened again and again. I would much rather spend two or three hours selecting twelve arrow shafts that hardly need any straightening at all.

Initially the shafts should be about three feet long and no more than half an inch in diameter, including the bark. Look for shafts that are as near as possible to the diameter of your finished product. You shouldn't have to do much more than shave off the bark and smooth it down. Most finished shafts will be somewhere between one-fourth and three-eighths inch in diameter, depending on the weight and strength of the wood: the lighter the wood, the thicker the shaft.

I suggest cutting a dozen to twenty shafts and making all the arrows at the same time. With the exception of reed and cane, these should be arranged in a bundle that is tightly and evenly wrapped with cordage and seasoned for a year, just like your bowstaves. Wrapping helps to keep the shafts straight as they are seasoning. About every two months I unwrap the bundle, jumble the shafts, and rewrap them so that all sides are equally seasoned. As for reed and cane, they should be

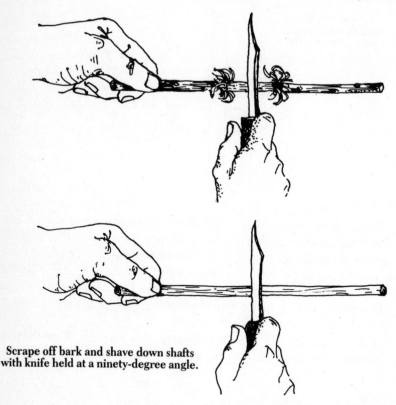

Scrape off bark and shave down shafts with knife held at a ninety-degree angle.

collected in the fall, just after they turn brown, and allowed to dry singly on a flat surface.

After the shafts have seasoned, shave off the bark the same way you did from the bowstave (scraping, not carving). This is best done by turning the shaft on your thigh with one hand as you scrape with a knife or thumb scraper held at a ninety-degree angle to the wood.

Smooth the shafts down to between one-fourth to three-eighths inch in diameter, using a coarse rock, grooved sanding block, or sandpaper. Gradually work on down to a fine grip sandpaper, emery cloth, or fine abrading rock. If you use sandpaper, wrap it on very tightly and close it with your fingers; then force the shaft back and forth through it. Be sure to pull the shaft all the way through so you sand the ends, too.

To make sure my arrows are of uniform thickness, I push the shafts through a series of different sized holes drilled in a bone or piece of wood. When I can pass the entire shaft through the largest hole, I continue to shave and sand; then I go to a smaller one. When I can pass the shaft through a hole just barely larger than the diameter I want, it is finished. Then I grease it with rendered fat and heat it by the fire to drive in the oils.

The next step is to cut the shafts to size. For a five-foot bow, arrows are cut and measured from the tip of the fingers to the armpit, which is usually about thirty inches. For a shorter bow, arrows will be about twenty-five inches. The exact length should be determined by the draw of your bow.

Smoothing Shaft With an Abrading Stone

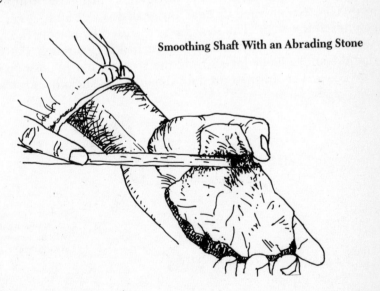

The safest way to make these cuts is to use your knifeblade like a lathe, turning the shaft as you wear a groove into it. Then, taking the shaft in both hands on opposite sides of the groove, move it back and forth until it cracks off evenly. The same is done with reed shafts, by moving both ends in a circular motion. (For further information on reed shafts, see *Reed Arrows*, page 150.)

Straightening Shafts. The shafts are straightened simply by heating them over hot coals and bending and holding them at the curves until they cool. Work on the major bends first, then go to the smaller ones. Warm the shaft until it is hot to the touch, then place your thumbs on the bend and pull in the opposite direction with flared fingers. Cool the shaft down by blowing on it and swishing it through the air as you continue to hold the bend.

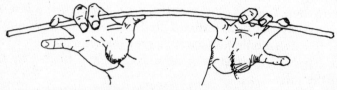

Use hands to straighten gentle curves.

Generally gentle bends can be taken out by heating the arrow and draping it over a log or a thigh. Smaller, more stubborn bends often require the use of an arrow straightener, made by drilling a hole through a piece of wood or deer antler that is just a bit larger than the diameter of the shaft. The straightener is used by placing the hot arrow (sometimes wetted with saliva before heating) through the hole and using the antler like a lever. Hold the straightener in position until the arrow cools. The finished product should be as smooth and straight as a store-bought shaft.

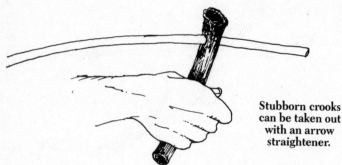

Stubborn crooks can be taken out with an arrow straightener.

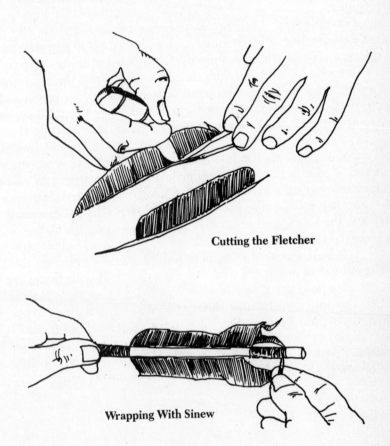

Cutting the Fletcher

Wrapping With Sinew

Fletching. The next step is to put the feathers on the shafts. Most native Americans, who looked to nature as their guide, fletched their arrows so they would be shot in the same direction they had grown— that is, sunwards. A few tribes, on the other hand, liked to shoot their arrows "downwards," the way the lightning travels, so they fletched their shafts at the top.

I find that six-inch feathers, trimmed to half an inch in height, are ideal. Turkey tail feathers are the best, though any tail feather from a medium to large bird will do. You can also use wing feathers, though they will make the arrow spin and slow it down somewhat. Just be sure to take them from the same wing so the arrow spins freely in one direction.

Carefully cut three fletches, either by slicing along the median line of each feather or by cutting into the shaft slightly and stripping off a length of feather with your fingers. Crop the feather so that you have about an inch of quill extending on either end. Trim the feathers to the proper height so they are all the same size and shape. (I prefer a straight fletch, with the feathers cut evenly across the top, though fletches with descending widths and rounded ends are also quite popular.)

Now, with the fingers of one hand, carefully hold the exposed quill ends in place on the inside end of the shaft and wrap them tight with sinew. Some people temporarily hold the feathers in place with pitch or diluted hide glue to free the hands. The sinew is prepared by pounding it into threads, then wetting it with saliva. Beginning at the base, it is then wrapped on evenly and snugly along the entire length of the exposed quills.

The best way to do this is to hold the sinew between the thumb and forefinger of one hand while turning the shaft with the other hand. You can also hold the sinew in your mouth and turn the shaft with both hands. The saliva and sinew mixture forms its own glue. Sinew wrapping is almost transparent, lies close to the shaft, and tightens as it dries. Alternatively, you can use artificial sinew (waxed cotton thread) by securing it first with half hitches on both ends and adding a thin coat of glue or pitch either before or after the wrapping is done.

**Nocking the Shaft
With a Stone Edge**

The back of the fletching is wrapped up to the base of the nock location. A U- or V-shaped nock is then abraded into the base of the shaft with a scraper or a small rattail file, all the way up to the sinew wrapping. This way the nock is supported by the sinew, and the string won't split the shaft when the arrow is shot. Be sure to file the nock in a position that will allow two feathers to pass the bow side evenly in flight. That is, if you imagine the feathers forming a triangle, the nock must be parallel to one of the bases of the triangle.

Attaching the Head

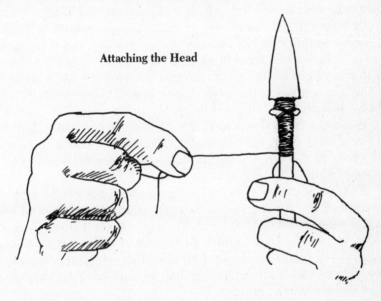

Arrowheads. Arrow tips vary with location, laws, and terrain. A survival arrowhead can be made of bone, antler, stone, metal, or fire-hardened wood. In New Jersey, where I live, it's illegal to hunt with anything other than a razor steel broadhead. Steel heads are easily cut out with a jigsaw and filed to sharpness. Take care to make a base that's large enough so the arrowhead can be firmly wrapped onto the shaft but small enough so it doesn't weight down the tip of the arrow.

Bone arrowheads can be almost as deadly as steel broadheads. Stone arrowheads can be made from flint, chert, jasper, quartz, obsidian, and even glass. These are just as deadly as steel when made in the proper way. For an explanation of bone working and arrowhead knapping, see *Stone and Bone*, page 101.

Always secure the arrowhead last, and always put it on so the wings are parallel with the nock. If you secure it at right angles to the nock, the arrow will be much harder to aim because the wings will flare out in your field of vision. If the arrowhead is vertically oriented, you can aim right off the sharp edge of one of the wings. Another advantage for short-range hunting (up to thirty feet) is that the ribs of most standing game animals are vertical, which gives the arrowhead a much better chance of penetrating past the rib cage. Beyond thirty feet the arrow will begin to spin somewhat, which will reorient the head.

In attaching the head, try to make it seem a part of the shaft. What you're working for is an elegant fusion of wood and bone, or wood and stone. Choose the arrowhead with care. Make sure it's neither too heavy nor too light for the shaft. Do this by tying the head on first and balancing the arrow across your index finger. The point of balance should be just about in the middle.

Hone the head down so that it fits snugly and naturally into the shaft. For a solid shaft, cut or abrade a notch that's just deep enough so that the base of the arrowhead fits snugly against solid wood. For reed shafts, see below.

Finally, put some glue into the groove and lash on the arrowhead with prepared or artificial sinew. Start by taking a few tight wraps just below the wings to secure the head. Then take several alternating diagonal wraps across the base. Finish by wrapping and knotting the sinew below the base. You can also add pitch or hide glue for extra security. Finally, shave or sand the tip of the shaft so it blends smoothly with the angle of the arrowhead wings.

Reed Arrows. Reed shafts make the lightest and fastest arrows, bar none. Fine arrows are made with reeds in much the same way as with solid wood, with a few exceptions. One difference is that reed shafts cannot be shaved down. For this reason, better balance is usually obtained by fitting the arrowhead onto the smaller end of the shaft.

Reed shafts are also very fragile. After the outer sheath is taken off, they should be straightened only at the joints, not in the hollow part of the shaft. They also require hardwood inserts for the nock and the head. This helps to keep the shaft from splitting, but it's not necessary for hunting small game.

The back end of the shaft is cut about half an inch from the joint. Here, a small wooden plug is inserted to receive the nock. This plug can be smeared with glue and wrapped snugly in place with sinew. The plug should fit snugly and you should start wrapping at the end of the shaft to prevent it from popping out.

A similar procedure is used on the front end. This end is cut about two inches from the joint and fitted with a four- to five-inch hardwood plug, which in turn is grooved and fitted with the arrowhead. You can also make a cylindrical arrowhead out of wood and fireharden it by turning it over hot coals before inserting it into the shaft.

Forward inserts must fit snugly against the joint of the shaft and be firmly secured with sinew wrapping. For best results, make the plug a little thick and bevel it at the base so that it slightly cracks the end of the shaft as it's inserted. Then, wrapping from the end of the shaft toward the joint, lash the shaft tightly against the plug and tie off the sinew.

Accessories

Before you take your bow and arrow hunting, be sure you've got the necessary accessories. These include an extra bowstring, extra arrows, and an arrow repair kit consisting of straighteners, abraders, sinew, pitch, hide glue, and a few extra feathers and arrowheads. All these things can be kept in a quiver—a rawhide, birch bark, or wooden tube that is nailed around a softwood base and then sewn onto a strap and slung comfortably over the shoulder. It's also a good idea to make a protective case out of buckskin or blanket material for storing and carrying your bow when you're not using it.

Close-Range Hunting

There is little I can say about modern, long-range bow hunting that is not covered in many other books, with the exception of stalking and camouflage (see *Blend and Flow*, page 193). However, close-range hunting is another matter. Close-range hunting is both an art form and a science, and it is here that you really begin to feel the bow and arrow as an extension of yourself. With practice, the bow may come to feel almost like another limb. You will know its strengths and weaknesses, and be able to move and hide with it so skillfully that you and your weapon become almost invisible against the landscape.

Accurate shooting is a matter of practice, but there are some general points to keep in mind. Unlike long-range hunting, in which you lift the bow grip to eye level and sight along the arrow shaft, short-range hunting requires the shot to be taken from lower down—usually from the chest, diaphragm, or abdomen. Also, the bow is often held diagonally across the body instead of vertically. Depending on the ter-

**The hunter must blend and flow with
the landscape while carrying the bow.**

rain and the circumstances, you may find yourself in some unusual positions, and you have to be ready to take the shot from wherever you can with as little movement as possible.

In short-range hunting, the shot is often taken from the chest or lower.

Also, rather than pulling the string back to the cheek as is the custom with long-range bow hunting, the bowstring is most often held stationary against the chest or belly with one hand while the bow is pushed out with the other to full draw. You must do this with the utmost care and caution, just as though you were a heron stalking a fish or a frog. Ever so slowly the bow is pushed out as you gauge the distance and range. Then at the crucial moment comes the strike. The arrow is sent to its mark just as suddenly as the heron's beak.

For successful hunting, nothing pays greater dividends than knowing the animal. I don't mean just book knowledge, but practical, down-to-earth experience in the field. You must study the animal at close range. That means tracking, stalking, and living with it until you know it like a member of your own family. As you do this, your hunting skill will increase exponentially, and so will your love and appreciation for the animal itself.

The next step is choosing an area. Most likely this will be a place where the animal appears at a particular time each day on its way to drink or feed. That is where you will lie in wait. Many modern deer hunters like to climb trees and stand on platforms attached to the trunk or branches. This works, but nature provides many other comfortable spots from which to survey the landscape undetected. The real challenge is to seek out these areas and make the hunt as natural as possible.

There are basically two options in short-range hunting: ambush and stalking. In ambush hunting, you and your bow essentially disappear. You become so blended with the landscape that nothing appears unnatural or out of place. If you are skillful and crafty, the animal will pass within feet of your hidden form. If you're stalking an animal, the hunt becomes a game of wits. It may last hours or even days, and it may involve endless moves and countermoves. To track and stalk an animal to within very short range (say, ten feet or less) takes such an intricate combination of skill and knowledge that it is exhilarating even to think about it.

There is another element to the hunt that is just as important as all the others combined. That is reverence and respect for the animal that is being hunted. The native Americans believe (and so do I) that in a truly sacred hunt, the animal's death is not so much a killing by the hunter as a sacrifice by the animal. They believe that the hunter, through his prayers and dedication, becomes so attuned to the animal's spirit that it finally gives itself to him in honor of his skill and devotion.

This is a difficult concept for many modern people to understand. It goes against the grain of our upbringing in so many ways. Yet time and again I have felt the flow of a special spirit, moving back and forth between me and the animal I was hunting. Time and again I have felt that awesome gift of a life given up, and the call to honor the sacrifice by using the animal well. This is the glue that holds the hunt together and makes it more than sport. This is the higher purpose of the hunter and of all the entities that go into the making of a finely crafted bow and arrow.

Note: For further reading on the bow and arrow, I recommend *American Indian Archery* by Reginald and Gladys Laubin, published by the University of Oklahoma Press. It includes a wealth of historical information, as well as illustrations of bows and bowmaking methods used by various North American tribes.

For further information on tracking, stalking, camouflage, and close range hunting, consult *Blend and Flow*, page 193, and *Tom Brown's Field Guide to Nature Observation and Tracking*.

8
HIDE AND HAIR

The number of useful items that were once made from the hides of animals is staggering. To name a few, it included leggings, skirts, breechclouts, moccasins, hats, cordage, shields, shelter covering, boxes, pots, and bowls. Probably no other animal was so revered among the native Americans for its hide than the deer, and no other material was so prized for its softness and durability than brain-tanned buckskin.

The tanning process that produces buckskin is a long and difficult one. It doesn't require much knowledge, but it does take a great deal of dedication and elbow grease. In my mind, though, there is almost nothing to compare with the product that comes from the procedure and the feeling of satisfaction that the work produces.

Mainly for this reason, this chapter will concentrate on brain tanning—the process of making deer hides into buckskin. But the deer is not the only animal whose hide can be tanned. With slight variations, this same process can be used for almost any other kind of hide, from elk, moose, and cow to rabbit, weasel, and mouse.

The Hide

Even if you aren't a hunter or trapper, there are lots of ways you can get hides. One of the surest is to purchase one from a commercial hide buyer. One disadvantage here, besides the cost, is that if you order through the mail you don't know for sure what you're getting. Sometimes the animals have been carelessly skinned and the hides are badly scored with knife marks. For this reason, I would recommend picking the hide out yourself. The same goes for meat processors and local stores, which also may be good hide sources. In fact, it would be wonderful if people began to really make use of the millions of animals killed every year by the food industry. Often all that is saved is the meat, and the rest goes to waste.

Another alternative is to take a tour of nearby ranches and farms and get the word out that you're looking for hides. Best of all is to keep your eyes open during the hunting season. You'd be surprised how many hunters throw their hides out because they don't want to be bothered with them. Let's face it, not many people are going to take the time and trouble to tan a hide into buckskin. But that's partly because they don't know what a rewarding experience it can be.

You'd also be surprised how many deer are hit by automobiles every year. Most of these hides go to waste just because state regulations usually forbid any tampering with road kills. But the law has its human side, too. You may be able to get a hide by explaining to the authorities what you are going to do with it and what a shame it would be to have it go to waste. Many of the deer and other animals whose hides have been tanned on my farm were road kills. Those deaths would have been totally meaningless were it not for the fact that somebody cared enough about the animal to take off its coat and turn it into a fine piece of clothing. I believe this honors the animal and puts its death to some useful purpose.

Skinning. If you kill the animal yourself, be careful to skin it in such a way that you don't damage the hide. One of the best ways to skin a deer is to hang it up with a rope by the neck and work the hide off as though you were pulling off a coat, top to bottom.

First, with a small, sharp knife, make an incision above the anus just under the skin. Don't poke all the way through into the body cavity! Just "unzip" the hide by slicing neatly under the skin all the way up to the throat. Next, cut around each of the legs just above the hooves and slice up the inside of each leg to the median cut. Finally, cut around the base of the rope at the neck and begin to peel away the hide from the neck down.

Now, this is where most people make their biggest and most costly mistake. They use a sharp knife to take the hide off. By hacking and slicing, they pull the hide off with thick chunks of meat and fat clinging to it and even score the hide unmercifully with knife marks. Some of these slices go almost all the way through the hide, which ruins that part for any kind of clothing. This is totally unnecessary.

In fact, there is usually no need to use your knife at all after the incisions are made. Most of the hide can be taken off by "fisting"—that is, through a combination of pulling on the hide and working your fingers and hands against the membrane that connects it to the carcass. It's very much like pulling off a tight-fitting coat. Best is to start with the back of the neck and work down from there, peeling and pulling, working your fist and fingers between the hide and the carcass as you go. Take your time, and avoid pulling off any meat. This will save you a lot of trouble later on during the fleshing process.

As I skin an animal, I also like to talk to it, much as I talk to a plant or anything else I'm harvesting or making into a garment or implement. In a physical sense the animal is dead, but to me its spirit is still very much alive. Whether or not the animal was killed by a bullet, an arrowhead, or a car makes no difference. It still deserves my respect, and its spirit deserves some reassurance that I will use its gifts in a good way.

So I talk to the deer as I take off its coat. I thank it for the gift of warmth and protection it is giving me. I feel the hide. Holding it in my hands gives me a luxurious, reassuring sensation. I feel the thickness of the dermal layers and imagine all the fibers that are going to be fluffed up in the tanning process. I feel the layer of thick, hollow, insulating hair that has kept the deer warm in the coldest of winter snows, and I am awed by its beauty and utility.

"Deer," I think, "your kind is lucky to have such a beautiful coat to carry on your back and belly all your days. We people are but paupers when it comes to having the things we need to survive. It is too bad we have to take so many things from you and your brothers and sisters, but we have no choice. We are poor survival machines. We rely on you for our warmth and well-being. Deer, I apologize for taking your coat, but I will make it a thing of beauty. And a part of you will live on with me as I bundle myself inside it."

Though "fisting" works, it is a little barbaric and rough on the hide. Most animal skins are tough, but all that pushing and pulling can stretch them, too. If you're not careful, you can rip a hole in a smaller hide. To avoid this, I have a variety of tools I use to separate the hide from the carcass without actually cutting.

One of these is a rounded, wood or bone knife that is not sharp enough to score the hide. It's about as sharp as a letter opener, but without any points. Instead of ripping the subcutaneous membrane as you would with fisting, the dull knife cuts and wedges through the connective tissue, allowing the hide to separate more easily. Another simple tool is a stick with a flattened, beveled, or dovetailed end that wedges easily between hide and carcass. Such a tool can also be made out of a thick piece of leg bone. It's safest and most effective if it's carved into a rounded, spatula shape so there are no points that might puncture the hide.

One of the best tools for skinning is a scapula bone that has been beveled on both sides and furnished with a sturdy handle made from a split stick that's lashed on with rawhide (see *Hafting Tools*, page 114). The scapula's wide, rounded end lets it work on a large surface area so that often the skin just seems to roll off the animal's back. It's a good idea to have several of these from different-sized animals so you can use them on different areas. You can use different-sized scapulas for all kinds of animal hides, from elephants all the way down to chipmunks.

Finally, for close, tight areas such as those around the legs, a thumb skinner is most useful. This is a piece of bone or stone that's shaped like a half moon and held between the thumb and fingers. It's much like a thumb scraper (often used for smoothing arrow shafts and other fine work), but it's quite dull. Even a quarter-sized skinner can be an excellent way of getting some of the tight, difficult parts of the hide to loosen and break free.

After the animal is skinned, quickly wash the hide in water, rinsing off any blood or other liquids that have adhered to it. Then, if the animal has been skinned by someone else, pull off as much fat and meat as you can. If you've skinned it properly, there should be no meat at all and only the thin film of fat that normally comes off with the bottom, or hypodermal, layer.

Now you have to decide whether you want to leave the hair on or take it off. If you're working with deer, elk, or any other animal that has a tendency to shed, I would recommend taking the hair off because eventually it's going to come off, anyway. If you're working with fox, mink, otter, cougar, or some of the other richly furred animals, the hide is more attractive and valuable if you leave the fur on (see *Tanning With Hair On*, page 174).

Soaking

If you decide to take the hair off, soak the hide in water for a day or more to loosen up the hair follicles. If possible, use the running water in a stream and secure the hide so the hairs point upstream. This way the current will help to loosen the hairs as it flows through them. If you put the hide in pond or lake, make sure the water is clear and free of microbes. The microbes in warm, stagnant water are plentiful, and they love to chew on hides. More than a day or so and the little beasts will probably begin eating away at the dermis, or inner layer, and after a few days the hide will be ruined.

The same goes for high temperatures. Never soak a hide in hot water unless you want to wear buckskin rags. Every body of water and every hide is a little different, so you'll just have to test the hide from time to time by pulling on the hairs. If you can't pull the hairs out even with real effort, leave the hide in the water a day longer. If the hair comes out too easily, you may have left it in a little too long.

Usually a two- to four-day soak is all that's required. A good variation on the soak is to immerse the hide in water that contains tannic acid, as this actually causes a chemical change in the fibers that makes them fluff up more. The Pine Barrens streams where I hold many of my classes contain quite a bit of tannic acid, and I always notice a difference in the quality of the hide when it's soaked in this water. You can get the same effect by soaking the hide in a bucket of water that's been treated with a shot glass or so of tannic acid, but if so, you should change the water frequently—two to four times a day, depending on the heat.

If there's still a lot of fat on the hide, you can loosen it up by soaking the hide for a couple of hours in a soapy solution. If you're somewhere close to civilization, you might want to use a mild soap like Ivory Liquid or baby shampoo (a capful in two gallons, well agitated, works well). In the wilderness I often pound up meadowsweet or yucca roots or mix a palmful of white wood ash with about two gallons of water. Even acorn boilings help to get rid of the blood and loosen up the fat, as they also contain quite a bit of tannic acid.

Fleshing

Now it's time to work all the excess fat and flesh off the under side of the hide. This is done by scraping the hide with a fleshing tool, which can be easily made from an old car spring—that is, one of the flat "leaf springs" used in the suspension system. You can get these at any junkyard, probably for free. Just file one edge flat so it forms a ninety-degree angle, and attach handles to both ends. You can also use a rock, rib bone, or any number of other things as fleshing tools, as long as the object has a flat, squared edge. This way you can put on all the pressure you want without breaking or scoring any of the dermis fibers.

The fleshing tool is held in both hands and pulled at a ninety-degree angle to the hide. In other words, there's no sharp angle biting into the hide. The fat and flesh residue is just "bulldozed" off the surface of the hide through pressure applied in short, even strokes. Another tool that works especially well for tough spots is a hoe-shaped blade with little square teeth filed into it.

For fleshing, the hide must always be put on a rounded surface. Best is a fleshing beam, which can be anything from a smooth, rounded log to a specially built hardwood beam supported on two legs. Ideal is a moderately hard log that's nice and rounded with a somewhat bullet-shaped nose. The hide can be safely stretched over the end of this to hold it in position while you scrape it with the flat implement. It can also be easily moved to different positions as you work on other areas.

**The Fleshing Beam
and Its Proper Use**

Be sure to do the fleshing with the hair still on. It acts as a
protective padding between the hide and the fleshing beam. If you plan
to leave the hair on the hide, make sure you scrape with the angle of the
hair follicles, not against them. That is, on a fur-bearing animal such as a
fox, you'll flesh from the tail to the head. That keeps the root hairs firmly
in place (see *Tanning With Hair On*, page 174). On the other hand, if
you plan to take the hair off, you can loosen it by fleshing from head to
tail.

As you do the fleshing, get familiar with the hide itself. Notice
any weak spots, bullet holes, score marks, and the like. Scrape at right
angles to any score marks, and take short, careful strokes around any
holes. Remember where these spots are, because you'll have to be
careful of them later. Even the scars the animal has picked up from
barbed wire fences or the bites of predators have probably bitten into
the dermis layer and pulled out hunks of hide. All these scars indicate

weak spots, and you'll have to use caution not to widen them or punch through the hide.

As you work, get a feel for the implement and how it slides over the hide. Keep the hide scraper clean and well honed. Feel it pulling off the white and yellow layers. All the fat and grease must come off. Feel the hide from time to time. If it's greasy anywhere, there's still some fat on it. Areas that are well cleaned will have some slight friction to them.

You may run into some problem areas while fleshing. Some of the more difficult spots are often on the back and belly—that is, in the middle and on the edges of your hide where lots of fat tends to accumulate. Sometimes it's easier to work these areas with a smaller tool such as a skinning knife or even a jacknife, again held at a ninety-degree angle. But be very careful not to slice into the hide.

Fleshing takes a lot of elbow grease. After a while the hide may begin to dry out and the fat and flesh will cling more tightly. If this happens, simply resoak the hide until it loosens up and go back to work.

Another necessity to this art is an indifference to grease and grime and certain raunchy smells that may be emanating from the hide. I don't know what to say about this except that you'll get used to it. Remember, the hide in its raw state is not what it will be when you get done with it. Part of the satisfaction in the finished product is the realization of how much of your own energy and elbow grease—your own "medicine," so to speak—has gone into the hide, and how much you and the hide have transformed each other in the process.

When the fleshing is done, soak the hide overnight, preferably in a soapy, tannic acid, or white wood ash solution. This helps to cut grease and loosen the hair follicles to make pulling the hair much easier. Once the hide has been rinsed, it will dry with a hard, somewhat shiny cuticle on the flesh side. This is very rough rawhide. You can use this for a number of durable items, from drumheads to strong lashing cord (see *Uses for Rawhide*, page 167). However, if you're planning to make buckskin, you'll either want to continue with the tanning process or store the hide in a cool, dry place.

If it's wintertime, I usually store my hides and work on them in the spring when the weather warms up. Some people like to salt their hides before they hang them up. Personally, I prefer to treat them with sodium borate (Borax), if it's available. This I rub into the fur and also onto the flesh side. Many insects and animals are attracted to salted hides, but there aren't many I know of that go for sodium borate.

For storage, rough rawhide should be tacked out in the open air.

Inside your earthshelter, unfinished hides can be used as temporary wall insulation and flooring until you get around to tanning them. You can even sleep on them. Just be sure to shake them out daily and keep the bugs away with cedar smoke or some other natural repellent.

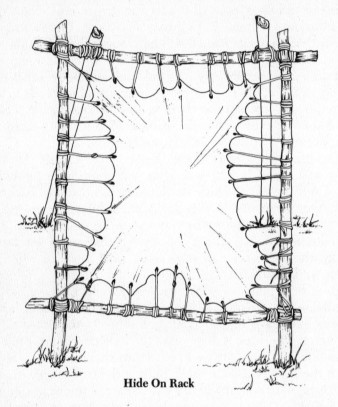

Hide On Rack

Racking and De-Hairing

To continue with the tanning process, I almost always string my hides up on a sturdy rack. It's possible to stake a hide to the ground. The Indian women did this with thick buffalo hides. But you'll notice there aren't many people tanning buffalo hides these days. And with most hides, it's just too easy to punch through when they're flat on the ground. It's also messier and more difficult. So unless the hide is quite small (for example, a rabbit or weasel hide that can be worked right in your hands), do yourself a favor and rack it out. This keeps it from getting soiled, excess hair and cuticle fall off the working surface, and you can move the hide to another location anytime you want.

A hide rack should be sturdy, convenient, and slightly larger than the stretched-out hide. For most deer hides I use a rack that is four to six feet square, made of two- to three-inch-diameter poles that are notched and bound tightly near the ends with square lashes. Sometimes I use two conveniently spaced trees with two sturdy poles lashed horizontally to complete the rectangle. All racks should have some means of support, whether fitted with diagonal "legs" that reach out from behind, lashed to a large tripod, or even leaned back against a building or tree. Outside, in most cases, they should be faced in a southeasterly direction to take best advantage of the sun.

The hide can be attached to the rack with any strong cordage. Rawhide is excellent in a survival situation. Otherwise I find baling twine the best because it's cheap, plentiful, and doesn't give much under pressure. First, with a sharp knife, slice one-inch slits every three or four inches all the way around the edge of the hide. These should be evenly spaced, about a quarter inch in, and parallel to the edge of the hide. One-inch slits pull more evenly on the hide than small holes, and the cordage has less tendency to rip them out.

Once the slits are made, tie a separate piece of four-foot cordage to each hole. One way to do this that also helps to distribute the tension is to double the cordage, slip the doubled end through the slit, and pull both ends through the loop. The ends are then secured to the hide rack with a square knot or shoelace knot. I recommend a knot that will be easy to untie later on, as it will be necessary to take the hide off and re-rack it twice during the tanning process.

Finally, just before you rack the hide, check it over for holes. If you find any, remove the hair around them and sew them up with a piece of sinew or other strong thread, using a simple overhand stitch or a baseball stitch.

To rack a deer hide or other "hair-off" hide, start by tying the neck to the top beam and the two hind legs to the lower beam to form a rough triangle. (For the "hair-on" process, rack the hide upside down so you won't be scraping against the grain of the hair follicles. Then rack the rest of the hide, alternating sides to keep it taut and even. When you're done, the hide should be firmly held between all four beams of the rack.

De-Hairing. The next step is to pull the hair off the hide. It's easier to take the hair off while the hide is still wet, so go to work on it right after you're done racking. You can pull off a lot of hair with your fingers, but some areas may require a more specialized tool, such as a custom-made hair pincer. This is especially true along the back and neck, where hair can be quite thick and tenacious.

One of the simplest pincers consists of two small pieces of wood that are tied onto your thumb and first three fingers. Another one is about the size of a large stapler. It's made by binding two flat, flexible pieces of wood or bone together at one end, then putting a small wedge between them so they open up about half an inch. These are used just like a giant pair of tweezers to get a better grip on the hairs. They can be made even more effective by carving a pattern of zigzag grooves into the surface of the wood. You'll probably be able to get most of the hair before you start scraping the epidermal layer of the hide. Any hair that's left over you can remove with the scraping tool.

Scraping

Before you begin scraping the cuticle, it's a good idea to let the hide dry about twenty-four hours on the rack. As it dries, it will shrink and tighten up. (This is why it's so important to have a sturdy rack.) For most of the scraping process you want the hide to be dry, so you can leave it in the open air until the cuticle layer becomes quite hard. The cuticle on both sides of the hide will dry with a dull, somewhat glassy sheen, and you'll probably be able to peel off large flakes.

The main object in scraping is to work that sheen off both sides until you get down to the soft, fluffy dermis fibers that are sandwiched in between. However, if there's still some hair on the hide, you should scrape off as much as you can before the hide dries.

Scraping tools, like fleshing tools, can be made from steel, bone, or stone. The most effective ones are made from pieces of leaf-spring iron that are rounded, beveled, and lashed or screwed onto sturdy handles. Likewise, scapulas, ribs, and cannon bones make good scrapers if they're cut and honed to the proper shape.

You can get by with one medium-sized tool, but it's a good idea to have scrapers of several different sizes to work different areas. I usually have a set of three, ranging from about one inch to two-and-a-half inches wide, plus a medium-sized scraper with blunt teeth filed into it for grabbing any leftover hairs. I also like to have a rock thumb scraper for finer work on delicate areas, as well as a few sanding stones of various sizes and grits (see *Stone and Bone*, page 101).

The blades of all scrapers should be beveled on the upper surface to about a sixty-degree angle. Don't put a knife edge on them or you will increase the danger of slicing through the hide. The idea is to scrape off

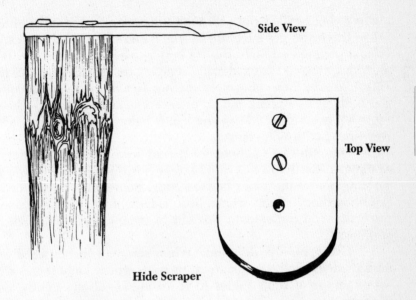

Side View

Top View

Hide Scraper

(not carve or slice) the cuticle layer, exposing the soft fibers below.

The tool must be used firmly but carefully. If you don't apply enough pressure, you won't get the cuticle off. Too much pressure will cause the tool to pop right through the hide.

Start out with gentle, downward strokes. Hold the handle firmly in one hand, and with your other hand on top of the scraper, pull the tool down in a slight arc. As you begin the stroke, the tool shouldn't even be touching the hide. As it comes down, it pushes in, slowly depressing the hide for a few inches as it shaves off a strip of cuticle. (Save the cuticle shavings—they make excellent hide glue and neat's-foot oil.) Gradually let up and come off the hide at the end of the stroke. The scraper is then lifted all the way up to start the next stroke. Don't scrape upwards, or you may damage the hide.

During the scraping process, keep the blade clean, and sharp. Above all, stay awake. Watch out for weak areas and don't stop in the middle of a stroke or you're likely to pop through the hide.

Every hide has its own personality. The way you work it depends on many factors, including the age and sex of the deer and even the time of year and conditions under which it was killed. Some hides are so tough there's almost nothing you can do to pop through them. Others are fragile and unpredictable. Once I was absentmindedly scraping on a deer hide when the scraper caught an edge, popped through, and made a hole so big I rolled right through the rack onto the other side!

Once you get a feel for it, you'll know how hard you can scrape without popping through. But it's better to be safe than sorry. It's also a good idea to work systematically, scraping one small area until it's fluffy white and soft, then going on to another area. I find the most convenient area is about a foot long and six or seven inches wide. Also when scraping, it's a good idea to keep the hands below the chin to avoid getting over-tired and losing control.

Sometimes it's not possible to get down to the buffy fibers just by scraping vertically. In this case I scrape horizontally, but I always pull in the direction of the tool's bottom edge. Sometimes I also encounter a "washboarding" effect, where little ridges of the hide are shaving off unevenly. You can smooth this out by scraping at right angles to the washboard.

Every hide is different. When scraping, be especially careful around edges, holes, weak areas, or spots where knife scores show up. Never scrape at right angles to score marks, always gently alongside

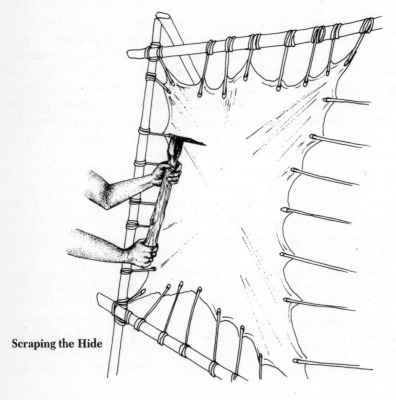

Scraping the Hide

them. And never scrape directly over holes. Go around them, preferably with a thumb scraper or sanding rock.

When you're done scraping the cuticle off the hair side, turn the rack over and work on the other side, the hypodermis. This side is a bit thinner and easier to work than the epidermis, but the same procedure applies.

When I'm done scraping an area, I like to finish up by sanding in quick, circular strokes with a fine-grade sandpaper or sanding rock. This helps to get rid of any excess cuticle and makes the hide feel almost like velvet or chamois, even before it's brained and buffed. This is fine-grade rawhide, and it can now be made into scores of durable items, from moccasin bottoms to storage boxes.

By the time you've finished scraping a hide on both sides, you will have had a major workout. This can take hours or even days of intermittent work. All I can say is that it's worth it. Many times while scraping or staking a hide, I find myself falling into a beautiful rhythm that connects me not only to the hide, but to the sun and the soil and the sounds and smells around me. I hear the birds and the laughing voices of children and friends. I feel the sun working on my own skin as I pour my effort and personal medicine into the hide in front of me.

Working a hide, like so many other earth skills, is a rare experience in modern society. It is a chance to forget the self, to let the mind drift and become part of the rhythm of nature. It is a time to forget time and destination. In a sense, it is a kind of meditation, and it can be every bit as profound a mental and spiritual exercise as sitting for an equal number of hours with your legs crossed and your eyes closed, staring at a candle or repeating a mantra.

Uses For Rawhide

When you're done with the scraping, you have a fine-grade rawhide that can have a number of uses. Although rough-grade rawhide (with the cuticle still on) is best for making things like lashing cords, shields, doorways, and boxes, fine-grade rawhide can also be used for these. One of the great advantages of rawhide is its tendency to shrink and dry hard. This quality makes it one of the best natural binding materials.

If you want to make rawhide cordage, the best way to cut a length of thong is to lay the rawhide out flat on a hard surface such as a log or table. Then, beginning on the outer perimeter, cut gradually in a spiral toward the center. This way, you'll be able to get a much longer piece of cordage than by cutting straight parallel strips. The width of the thong depends on how you plan to use it. A half-inch strip twenty feet long can be soaked, then twisted and stretched into a much longer and thinner cord. If you use it while it's wet, it will shrink and dry as hard as the strongest glue (see *Hafting Tools*, page 114).

Rawhide can also be cut to pattern and sewn into a variety of tough articles. It makes durable soles for footwear, excellent cooking and storage containers, useful shelter material (tarps, doors, rugs, roofing, etc.) and stiff but serviceable emergency clothing.

To make small rawhide containers, soak the material in water, shape it into a pouch, and fill it with wet sand. Wrap cordage around the mouth and allow it to dry. Gradually the rawhide will shrink to about half size. When it's dry, pour out the sand and use the container for whatever you want. You can make larger containers by pushing well-soaked patches of rawhide into the ground and filling them with sand or stones. When dry, such pots and pans can be used to hold either solids

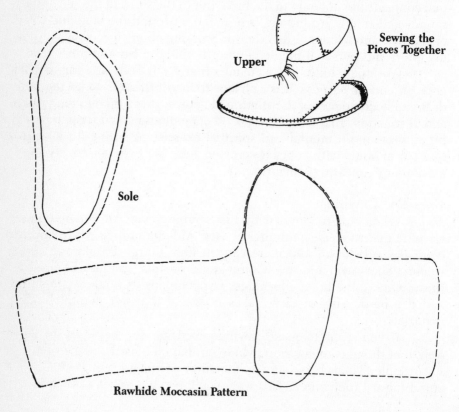

Upper

Sewing the
Pieces Together

Sole

Rawhide Moccasin Pattern

or liquids. You can even use them for boiling and cooking (see *Fire in Rocks*, page 81).

Braining

When the dry scrape is done, take the hide off the rack by untying all the individual strings. Leave the strings attached to the hide for easier restringing later on.

Now comes the magical part of the tanning process, the braining. This is the part that dramatically transforms the hide in the space of a few seconds. Actually brain tanning is a misnomer. It's not so much a chemical reaction as a physical conditioning. As the brains are absorbed by the hide, they add oils and enzymes to the dermis fibers that strengthen and allow them to fluff up into a softer material.

The old saying that every animal has enough brains to tan its own hide is generally true, but it's not necessary to use deer brains for a deer hide or cow brains for a cow hide. Usually about a half pound of brains is enough to tan the average deer hide. Better to have a little too much than not enough. If you haven't killed the animal yourself, you can get brains at a number of different places, including local markets and slaughterhouses. They can be kept indefinitely in a freezer.

If the brains are frozen, thaw them out. Then heat and gently mash them up until they are mildly warm and about the consistency of a gooey paste or pudding. Work them in your hands or a blender until there are no lumps, and try to take out any bone fragments that might have fallen in. Also, since brains have a way of smoothing out rough spots on the hands, I suggest wearing rubber gloves if you need your calluses.

If you like, you can use the brains just as they are. But you can also increase the life of the hide by adding a little bit of tannic acid solution or a few drops of neat's-foot oil (see page 139). I often do this because I find they help the brains penetrate the hide and add a little softness to it.

When the brains are ready, soak the hide in lukewarm water for about five minutes, then wring it out gently and lay it on a clean, flat surface. Now vigorously rub in the brains with your hands. Work them in well, rubbing in a circular motion until you're sure the entire hide has been completely permeated. When you're done with one side, flip the hide over and do the same on the other side.

As you do this, the hide will shrink and get very thin and translucent. It may even feel like a big piece of spaghetti. This is normal. Once the hide is well saturated, mix the rest of the brains in a bucket of water and immerse the hide in it. By this time the hide will be so small you can probably pack it inside a gallon coffee can. Let the hide sit in the brain-and-water mixture for about twenty-four hours, manipulating it occasionally to make sure the liquid soaks all the way through it. Then re-rack the hide.

Stake Stretching

Now comes the last great effort that will make your hide all it was meant to be. This is the stake stretching, or staking. Basically what you have to do now is push, pull, punch, and manipulate all parts of the hide without letup until it is soft and dry. What this does is stretch, loosen, and fluff the brain-impregnated dermis fibers, giving the buckskin the luxurious softness and pliability that makes it so special.

The duration of the staking process depends a lot on the weather. If it's a warm, sunny day, it may take only two or three hours. If it's cool or if you're inside, it can take up to a full day. Average staking time is from two to four hours. All things considered, it's best to get ready for the long haul and allow a full day. Staking is an almost nonstop process, and it's critical to the final appearance of the buckskin.

After the hide has soaked for a day, take it out of the brain-and-water mixture and wring it out. If it looks like a hot day, just wring it gently in your hands. (Otherwise it may dry too fast for you to get it well stretched.) On a cool day, wring the hide vigorously, either with your hands or by twisting it tightly with a stick. Don't wring so hard you tear the hide, but wring it hard enough so you won't be staking all day. You'll have to gauge this yourself. If you're outside, you can move the rack in and out of the sun to speed or slow the process.

Staking Tools. Any number of rounded objects are useful for staking. To get a feel for the hide, start by pushing against it with your hands and fists. Don't push too hard in the beginning. Just feel the hide stretching and telling you how much pressure to apply. Soon you will be able to push harder. As the hide dries you'll find you can lean all your weight into it with no danger of popping through.

Don't work too long with the hands and fists, or you'll rub them raw and tan your own hide. After you have a feel for it, go to a wooden staking tool. There are many smooth, rounded objects that are useful for this. I would suggest starting out with a wide tool at first, such as a handle on a canoe paddle, and later moving to a smaller staking tool such as a broomhandle.

As you get a feel for it, push harder against the hide. Really stretch and shove it. Push in and down with the tool, getting an easy but firm rhythm going, and lean into the hide so you're not doing all the work with your arms. Stake the hide continually until it's soft and dry. If it's drying fast, either work faster or put the hide in the shade. If it's drying slowly, work more gradually. Stay in tune with the sun and the wind and your inner voice.

When the hide is dry, it will no longer be cool to the touch. That's your indicator. Put your cheek to it. The best place to test is around the tail and back legs, since they tend to dry last. Work the entire hide, including the edges, until it is soft and pliable.

Staking puts tremendous pressure on the hide, and it's likely that one or more of your strings will pull through. As long as the hide doesn't get too loose, don't worry about it. If the hide stretches and gets too flabby, tighten up the loose strings and replace the ones that have pulled out.

One more word about pressure: The harder you stake, the more the hide will stretch and the thinner the buckskin will be. The less pressure, the thicker the material. Generally, if you want the hide for summer wear, stretch it more and keep adjusting the ties. For winter wear, stretch it less. The amount of pressure must also be geared to the size and strength of the hide.

Buffing

After the hide is staked and completely dry, take it off the rack and cut off the unusable parts around the edge. Now it's ready for buffing. Buffing is the process that loosens and fluffs up the fibers to increase their softness. I prefer to buff my hides by pulling them back and forth around a thick rope that's tied in two places to a pole or tree trunk. The coils and twists on the rope not only heat up the hide, but pull and tug at the fibers. This gives them that final, soft fluffiness.

Some people buff their hides around tree limbs or poles. This is all right as long as the wood has some abrasive texture to it. If not, all you'll do is heat up the hide. (I've seen some hides begin to smoke while being buffed, so don't go overboard!) By the time you're done with the first buffing, your buckskin should be as soft as a baby diaper.

Buffing the Hide

Smoking

After I've given the hide a thorough buffing, I like to smoke it and then buff it once more. Smoking the hide cures it, washes away the brain smell, and gives it some oils that help to preserve and make it water-resistant. It's best to use an outdoor fire for this, though when I was a boy I often smoked smaller hides right inside the chimney of my parent's house (much to their dismay—especially with the larger ones that backed the smoke up into the living room).

If you're smoking inside your earthshelter, the best setup is to make a fire in the pit and let it burn down until you've got a good bed of red-hot coals. Then on top of the coals place a good portion of wood chips that will produce a lot of smoke with no flames. I don't like to use green or pitchy wood, since it produces too much steam and sparks.

Different woods will turn the hide different colors. Cedar, for example, tends to smoke the hide a light amber, while hickory will turn

Smoking the Hide

it almost black. A loose rule is that the softwood chips tend to smoke a lighter color than the hardwoods. I find that burdock, cedar, and alder are good smoking woods, but you can also use things like yucca, sage, and even bark chips from various trees.

Before you start smoking in earnest, make sure your fire draft is working well, with the smoke flap turned into the wind. Then sew up the hide and suspend it in the shape of a cone just above the fire. The idea is to make a kind of chimney out of the hide so the smoke permeates every pore. Once it's set up, you can regulate the amount of smoke by opening and closing the draft holes on the outside of the shelter. Also be ready to spread out some of the wood chips if they really flame up.

There are many other ways to smoke a hide. A very effective method out in the open air is to make a dinner-plate-sized fire in a pit about two feet deep that's lined with rocks. Let this fire burn down to a good bed of coals, then punch a draft hole two to three inches wide down to the base from the direction the wind is blowing. Finally, ring the little fire with a section of hollow log or some other kind of tubing (small garbage-can cylinders work well) and wrap the end of the hide around the top of it to form a slightly open cone. Suspend the hide with a tripod. Then add your wood and let it smoke. You can control the amount of smoke and heat by adjusting a slab of bark on top of the draft hole. If the coals are going out, give it a little more air. If the fire flames up, close off the hole.

Smoke both sides of the hide for a total of several hours each, or until you're sure the smoke has permeated the entire hide. Then let the hide air out for a day.

Tanning With Hair On

Before I go into the making of various kinds of clothing, I want to explain more specifically how to tan a hide with the hair on. Most important is to keep the hair from falling out. For this reason, never soak the hide for a long period of time. Simply rinse it out and go through the fleshing process as described on page 159. However, do not flesh the hide against the grain of the hair follicles. Put the hide on the beam with the head of the animal up and scrape downwards, or vice versa. Once the fleshing is over, you should hardly have to touch the hair at all.

When you're done fleshing, rinse the hide again and tie it to the rack. Do not soak it overnight, and be sure to rack it with the hair pointed upwards (that is, head down and tail up). The scraping process is the same as before, except that you only scrape the flesh side—always with the grain of the hair.

For braining a "hair-on" hide, do not take the hide off the rack. Just pick up the whole rack and set it on four logs or other supports, hair side down. (This setup looks a little like a trampoline.) Then take a handful of water and dampen the flesh side of the hide. There is no need to soak it. Mix the brains into a paste as before, this time adding perhaps a shot glass full of tannic-acid water to make them penetrate more deeply. When the brains are ready, rub them in thoroughly, but only on the flesh side. Then put dampened towels or grasses over the hide for twenty-four hours. If it starts to dry out, just add a little more water.

The next day, stand the rack back up and lightly scrape off any brain residue. Then, using the methods and tools described above, stake the hide on the flesh side only until it is dry. Do not stake so hard that you rip or pull out the hair follicles. Finally, buff the hide thoroughly with sandpaper or a sanding rock to get it fluffy. Then smoke only the flesh side. When the smoking is done, you'll have a hide that's ready to be used as a bear rug, a coonskin cap, or whatever you want to make.

Buckskin Clothing

The best way to learn how to make buckskin clothing is to do it. Most of the native Americans were master tailors. As with their tools and weapons, however, you would never see them fitting a piece of buckskin with a tape measure or tracing patterns on the freshly tanned hide. They did almost everything by feel and by eyeing the person and the material.

Once you get a feel for buckskin, you can do this, too. Before long you'll be able to look at a hide even during the tanning process and begin to get visions of what it might be. You'll know almost instinctively where to cut to make a shirt front, a sleeve, or a pair of moccasins. This all comes with time and experience.

Buckskin likes to hang on your body the same way it does on the deer. For this reason, mark your patterns or make your cuts parallel to one axis or another of the animal. Buckskin that's cut on the skew tends to stretch more and doesn't wear as evenly. This approach also tends to waste material.

If you're making a shirt, you can start by folding one hide width-wise in half and draping it over your shoulders, then adding hides both front and back to complete the length. Sleeves can be made from sections that are cut from the leg and shoulder areas of the animal.

Another way of making a shirt is to use two hides—one for the front, one for the back, and the leftovers for the sleeves. Pants can easily be made by wrapping a hide around each leg, then cutting, patching, and sewing where necessary. In other words, you don't have to produce the best-tailored, form-fitting clothing. The idea is to make the act of tailoring an extension of the hide tanning itself—an exercise in feeling and connection with the hide and the animal that gave it.

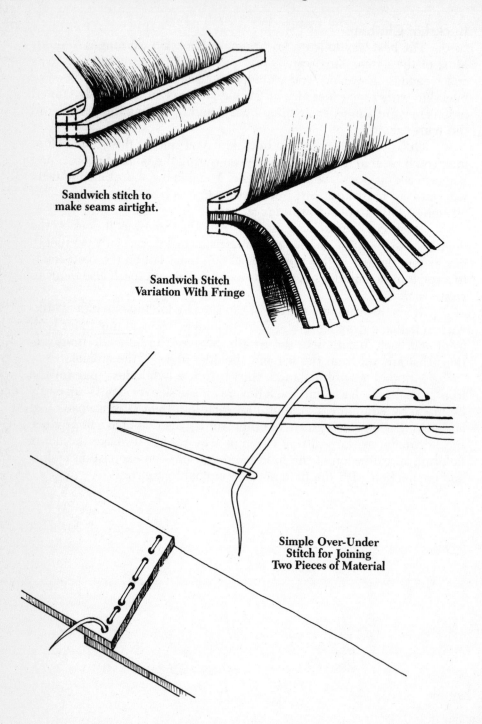

Sandwich stitch to
make seams airtight.

Sandwich Stitch
Variation With Fringe

Simple Over-Under
Stitch for Joining
Two Pieces of Material

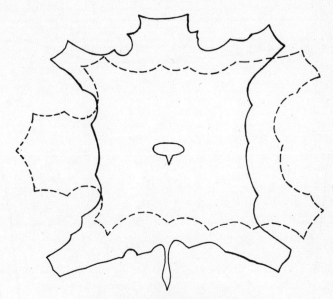

Two hides can make a large shirt,
with material to spare.

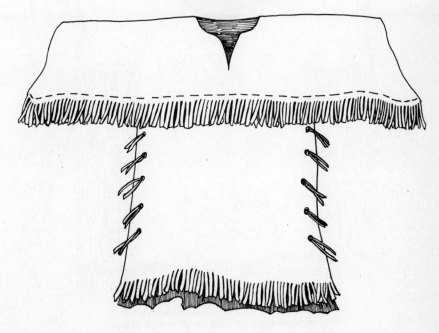

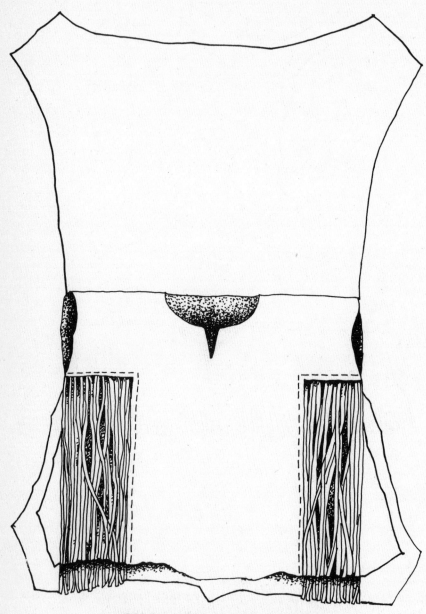

A folded hide can be made into a shirt.
Excess material can be used for sleeves or fringe.

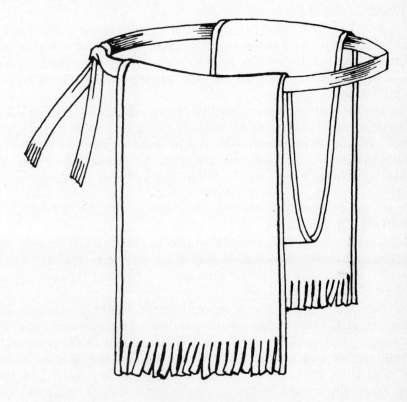

Belt and Breechclout

Again, ask the hide what it wants to be. Explore its size, shape, and texture. As you work it, think about what kind of clothing or article would seem most natural for it. Following are some basic patterns for some of the most common and popular kinds of clothing, plus illustrations that will suggest how you can cut and stitch them together using anything from rawhide strips to plant fibers.

Cleaning Buckskin

Buckskin can go a long time without being cleaned, partly because earth and soil blend so well with the natural color of the material. If the garment gets really raunchy, you can wash it in a mild soap. But this will remove some of the brain oils and smoke, which will reduce the hide's suppleness and water resistance. For this reason, I prefer to use a soap with a natural base.

Wet buckskin should be allowed to dry in a warm place, but not in direct sunlight. As it dries, every so often you should stretch it and buff it. Otherwise, it may dry hard in spots. The best procedure is to wait until it's almost dry, then give the whole garment a good buffing around a rope or branch.

If you find you've washed too many oils out of the buckskin, smoke it again. This will help to keep it soft and water-resistant. If you're wearing a wet garment, keep it on until it dries out to prevent shrinkage and then buff out the hard spots. With the proper care you should be able to keep your buckskin soft and serviceable for many years.

The Final Connection

Tanning hides and making your own clothing are complicated and time-consuming pursuits. It may seem at first that what you get from the process isn't worth the time and trouble. But the tanning and tailoring is only half the story.

The other half is the wearing. Just the feel of buckskin is like nothing else. It's one of the most natural materials you can wear. It's soft, supple, and so yielding it seems to hang on your body as naturally as it did on the deer. When I put on a buckskin shirt, I immediately relax, and I can feel a certain wildness come over my body. I believe there is actually a spirit in the material that most clothing items don't contain. I can't explain it, but it's a little like taking on the spirit of the deer, as well as some of the grace and elegance of its movement.

Through the shirt there is also a spiritual power that helps connect me to the earth. I feel the same sensation with moccasins. It goes beyond just the feel of the earth that comes up through the soles. There's almost a rootlike connection there that you don't get with Vibram soles or even sneakers. Anybody can tell the difference.

The native Americans believe very sincerely that the things that were made from an animal still possessed part of the spirit of that animal, and so do I. When I'm wearing my buckskin shirt on a cold day,

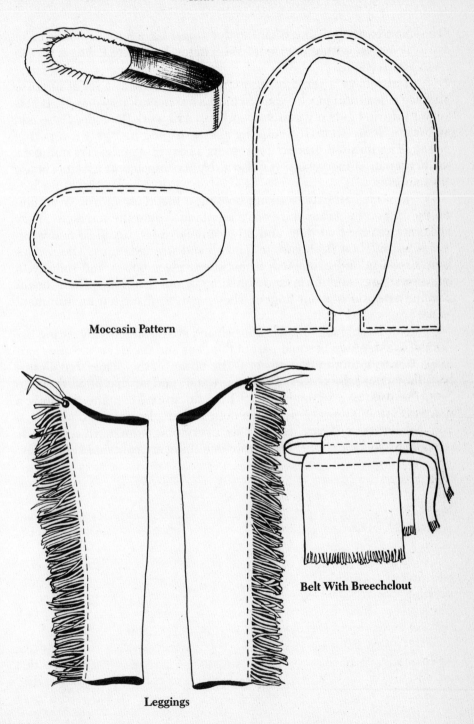

Moccasin Pattern

Leggings

Belt With Breechclout

I can more easily imagine what the deer must have felt like on a similar day as it moved within that hide. With natural clothing I always feel an earth connection.

I also have a sense of accomplishment. It takes a lot of effort to produce a beautiful piece of buckskin. It's a tremendous workout. It gets me outside and gets my muscles working. As I work the hide, I also add my own essence to it. In sculpting and reshaping it, I give it a part of myself. The finished product represents a beautiful union. It's not just a shirt or a pair of moccasins. It's also a relationship and a rich storehouse of memories.

For me, a rawhide thong holding a blade on an axe or an elegantly fringed buckskin shirt serve as windows into the universe. They make me realize that deer and other animals serve us in so many different ways. Their flesh goes to nourish and strengthen us. Their hides keep us warm and protected. Their sinew gives power and solidity to our bowstrings and form and function to our clothing. Their bones become extensions of our fingers, allowing us to shape our environment as we are shaped by it.

Not least of all, animals nourish and strengthen us. By eating the flesh of the deer we become part of the deer and the deer becomes part of us. Contributing to that power is the power of the plants that animal has nibbled and the power of the sun, water, and air that lifted them up from the soil. Everything we take from an animal contains not only a practical use but a precious connection. In a single piece of buckskin, if you look with your heart, you can see everything from earth and clouds to sunlight and starshine. For within one thing are contained all things.

9
EARTHENWARE

I am not surprised that among the North American Indians it was the woman who developed pottery making into such a fine art. To me, clay is like the nerve fibers and blood vessels of Earth Mother, and it is the woman who is in closest touch with the nurturing powers of the earth. While the raw material was being prepared and the pots being made, the clay became part of the very fiber of the woman's hands. She gathered it, pounded it, sifted it, washed it, kneaded it, molded it, fired it, smoked it, and decorated it. And in the process she mixed a part of herself and her tribal traditions into it. These were passed on to others not only through the sight and feel of the pottery, but through the love and power of the woman herself.

Pottery has many of the same uses as baskets. It is especially good for cooking and storing water and other liquids. But since clay can be molded into virtually any shape, its uses are nearly unlimited. Not only can you make it into earthenware pots, jugs, and cooking utensils, but also cups, dishes, pipes, and even ceramic toys and figurines.

Preparing the Clay
Deposits of clay are formed when fine silt suspended in water settles to the bottom and piles up over a long period of time. For this reason, most often you can find fairly good clay deposits along water-

ways such as riverbeds, streambeds, and ancient or dried-up lakebeds where the layers have been exposed by erosion. The best clay for earthenware pots is made up of fine particles that contain very little sand. (If it contains too much sand, it will crumble after it's dry.) When you squeeze the raw clay between your fingers, it should give a little but not crumble. When absolutely dry you should be able to crush it into a fine powder.

Crushing. Prepare the clay by first breaking it up into little bits and pieces and letting it dry completely in the sun. Then pound, grind, crumble, and scrape it against a clean, hard rock or other flat surface with a rock, mallet, club, or flat stick. Work it until it's the consistency of a fine flour.

Pounding the Clay

Sifting. Next, sift the clay to get out all the pebbles, sand grains, and impurities. The most efficient way to do this is to shake the flour through a sieve, screen, or fine-mesh basket. Another method that the Indians used was to throw the clay dust into the air from a flat basket or tray on a mildly windy day. The sand grains and other large particles would fall back into the tray, while the clay dust would settle downwind on a blanket or mat. Whichever method you use, it's a little like aerating the earth and honoring all the little particles that will soon materialize into a form of your own imagining.

Sifting the Clay

Tempering. Once you've prepared enough clay, put it in a wooden vessel and strengthen it by mixing in finely ground seashells, powdered eggshells, pottery fragments, or even crushed limestone or sand. A little of this tempering mixture goes a long way. A palmful for every quart of clay acts as a catalyst to help the clay harden and prevents it from cracking while it's being fired. But too much can cause the clay to crumble.

Kneading. When the mixture is just right, gradually add small amounts of water and work the clay in your fingers until it's about the consistency of stiff putty. Don't worry if the clay is too dry and tends to crack or crumble. You can always add more water. If the mixture gets too wet, add more powdered clay until it's a stiff consistency. Work in a handful at a time and smooth out all the lumps as you go. Also work out the bubbles; otherwise they will expand and cause the pottery to crack during the firing process.

Kneading

Every clay is a little bit different. Try to "read" the mixture with your hands as you knead it, calculating just how much water to add and beginning to formulate in your mind what you want to make. Work each handful of clay for five to ten minutes to make sure it's well mixed. When you have enough, you're ready to start shaping the container.

Making the Container

The kind of earthenware you make depends on your needs and preferences, but it's best to start with something simple like a small, flat-bottomed pot or cup. This way you can learn the basic principles, then go on to other projects with more confidence.

There are two simple methods of making containers. One is the "pinch" method, in which you take a lump of clay and mold it with your fingers into the desired shape. The other, which I will outline in detail here, is the coil method, where the walls of the container are formed by coils of clay stacked on top of each other. The pinch method is faster, but the coil method makes better pottery.

Forming the Base. The instructions for pots, bowls, and cups are very similar. Start by flattening a lump of clay into a pancake shape that's just big enough for the bottom of the vessel. Smooth this out with your fingers and round it carefully until it's a uniform thickness on all sides. A good thickness is a little less than a quarter inch for most cups;

somewhat thicker for pots and large containers. As you work, wet the clay with your fingers, rubbing it in a circular fashion.

Take plenty of time with the base, as this is the foundation. When you're done, allow the base to dry for about five minutes until it's no longer wet and gooey. Finally, score the inside edge with a stick or other instrument. This will help the coils to stick better once you begin forming the sides.

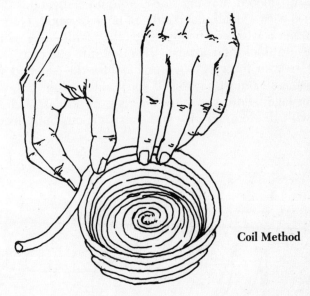

Coil Method

Coiling. Next, roll the wet clay into coils of the desired thickness between your palms and begin building up the sides. Make the coils as long as possible. Wrap them around the edge of the base either separately or in one continuous strand, splicing on new sections as you go upward. Make sure the base coil is wide enough to support the weight of the others, and smooth the coils with wet fingers as you go.

Wherever you want to widen the container, make each succeeding coil slightly longer and stagger it over the previous one. Where you want the vessel to be thinner, stagger the coils inward by making them shorter. Put score marks on top of each coil so the next layer will stick better. If your pottery is going to be constricted toward the top, smooth the inside while you can still get your hand in and out (see *Smoothing* below).

Toward the top of the pot you can add handles by making thick coils with sturdy bases and smoothing them into the surface of the vessel with plenty of clay. You can also add lids to open vessels. These can be made in much the same way as the base, from a single lump of clay. They fit more snugly if you also attach a thin ring of clay just inside the lip of the container.

Smoothing. The next step is to blend the coils into each other so they form smooth, flowing sides without corrugations. The best way to do this is with a small, flat paddle of bone or wood. Gently holding onto the container with one hand, scrape the paddle over its inner and outer surfaces until all the coils have blended into one consistency and thickness. Use your fingers, too. Get a feel for the clay and smooth it in the way that seems most natural. If you want a softer tool, wrap the paddle with cordage before you work the clay. As a final touch, make a watery paste with the leftover clay and smooth it on, both inside and out.

Smoothing Pot With Hand and Wooden Paddle

Decorating With Pronged Stick

Decorating. Finally, while the container is still wet, put whatever permanent designs you want on it. The Indians had several common methods of doing this. One was to score the pot with a many-pronged piece of wood or bone, which left a series of parallel lines that could be arranged in any way the artist wanted. Another was to use a similar tool to leave little "toothmarks" in various patterns. A third was to scrape the edge of a seashell back and forth on the clay surface, which left a curved zigzag pattern. And finally, sharp-pointed sticks and bones were used to draw either free-flowing or geometric designs. This was primarily an artistic touch, but I recommend it for two reasons. One is that it's a way of putting your own personal stamp into the clay. The other is that these marks help to transfer the heat during the drying and firing processes.

Firing. When the pottery has been shaped to your satisfaction, let it dry for up to two days. Then fire or bake it to make the clay set hard. There are several methods of firing. One is to find a sandy area and bury the pots from six inches to a foot beneath the ground, making sure you don't dent or damage them; then build a fire over the top and keep it going for about eight hours.

Firing Methods

Finished Pottery

A second method is to build a good-sized cooking fire and place the pots upside down above it on a green, wooden rack. This way the heat and flames lick upward, eventually reaching the entire pot. Finally there's the primitive kiln method, where you build a rock oven with a door on it and fire your pottery through heat from a fire below. If you use this method, leave the pottery inside for about six hours.

Firing is a tricky business. The pottery must be done but not overdone. If the fire isn't hot enough or you don't leave the pottery in long enough, it comes out weak. If you get the fire too hot or leave the pottery in too long, it comes out cracked. I can only give you general guidelines about this because every situation is a little different. But after a few trials, you'll get a sixth sense for how big a fire to make and just how long to leave the pots inside.

Finishing. After the pottery is fired, cool it overnight. The next day, take it out and flick it with your finger. If it has a nice ring to it, you know there are no cracks in it. If it doesn't ring, it's probably cracked somewhere. This doesn't mean you can't use it, but don't be surprised if it eventually breaks.

If the pottery rings nicely, I usually sand it down with some gritty material and then smoke it thoroughly inside and out over a fire burning with dried bark or some other smoke-producing fuel. This cures and waterproofs the pottery. You can also do this immediately after the firing, by throwing dried pith or some other smokey fuel on the coals and inside the pottery.

The final smoking is not absolutely necessary. If you have access to modern materials, you can buy glazes to take the place of the smoke and provide a final sheen, but you can also use pitch to help waterproof the pottery, just as for baskets. After that, you can paint or decorate it any way you like. If you don't smoke your pottery or put some kind of glaze on it, you can expect it to leak or "sweat" somewhat. But this oozing and evaporation can also help to cool the water.

Other Uses For Clay

There are lots of other things you can make with clay. To make a pipe, simply place two dowels at right angles to each other—one short and thick, the other long and thin—and cover them with a thickness of clay. If you fire this hot enough, the wood burns away or you can chip out the char, leaving a hollow bowl and pipestem. Flutes can be made in a similar manner, and beads can be made by firing tiny balls of clay strung on strands of cordage.

For a summary of the various pottery- and basket-making traditions of the native North Americans, as well as some excellent and colorful design ideas for these and other crafts, I highly recommend *North American Indian Arts*, by Andrew Hunter Whiteford, edited by Herbert S. Zim and illustrated by Owen Vernon Shaffer. This little *Golden Guide* is clearly and simply written, with beautiful illustrations.

Final Thoughts

Again, the making of the pottery is only half of the experience. When you drink water out of a clay mug or bowl, somehow it's very different from water taken from a plastic, aluminum, or glass container. Clay gives an earthiness to everything it contains. It's like drinking from the cupped hands of Earth Mother.

The feel of the cup and the flavor of the liquid also bring back precious memories. When I drink from such a container, I can see the place where I gathered the clay. I can remember the raccoon tracks I saw embedded in that special spot, and I can see the deer that came by while I was digging in the soil. I can recall the feel of what it was like to be alone beside that ancient riverbed, talking with layers of earth that were laid down with such patience and artistry over so many years.

As I drink from my clay cup, I can also recall the various stages in the making of it. In an instant I see how it all happened, from the pounding and sifting to the molding and firing. I think of how the clay was broken down under blows from my hammer, much as its particles were eroded from rocks so long ago. I see a connection between the airy particles falling onto my collecting blanket and the silt particles drifting to the bottom of an ancient lakebed. I know how the clay felt when it was washed by the waters of the earth because I, too, have washed it. I know that just as this clay was bent and shaped by my own hands, it was bent and shaped before by the hands of the Creator. And when I think of the final firing, I remember the heat and pressure in the bowels of the earth that mold and temper the great ocean basins and create the raw material for the sculpting of new landscapes.

In other words, working with clay is realizing the earth connection in both a symbolic and a real way. On a practical level, you realize just how important containers are in a full survival situation, and how you have made a piece of pottery that is an extension of yourself. It fulfills a need and brings you closer to the earth. On a symbolic level, it's like going through a little cosmic ritual in which you realize more fully just what clay is and how it connects with everything around it. That's when all the care and devotion you have put into this little chunk of earth come back to give you their greatest rewards.

10
BLEND AND FLOW

So many people ask me, "Tom, how can I get closer to nature? How can I begin to feel that ancient oneness you're talking about?" The answer is in your own heart. It is not necessary to make a bowl or a basket to feel your ancient connection to the earth. There are many other pathways. The one that you take should be uniquely your own. I can only point out some of the landmarks.

The most important of these is your mental attitude. We've only been "civilized" for a few hundred years, yet our bodies have the genetic memory of millions of years of living with nature. We all have the ancient harmony within us. The only thing that stands in the way is logic. When you can let go of the logical mind and allow yourself to feel your natural excitement and spontaneity, your body will begin to take over. Ancient memories will be reactivated, and you will begin to blend and flow as you were meant to.

There is an old Indian prayer that goes something like this: "Grandfather, grant me the ability to feel the same awe and inspiration in your little miracles that I do in your grand wonders." This is a tremendously important attitude to have. When you go out to enjoy nature, leave all expectations behind. Be spontaneous. Look at everything with equal curiosity. Go to the woods for the sheer enjoyment of experiencing *anything*. Then everything will be a source of wonder and enjoyment.

This will happen especially if you let go of time and destination. I don't care where it is—city, suburb, wilderness, or any place in between—time and destination destroy the eternal now that is the basis for living in harmony with the earth. What is the eternal now? It is having no past and no future. It is having only the present moment and enjoying it fully and completely.

Living in the now means many things. It means sitting by the fire in your earthshelter and listening to the crackle and hiss of the flames. It means feeling the warmth and light, hearing the katydids and cicadas beyond the circle of your hearth, and sensing the stars above the smoke flap. It's feeling all the other phenomena that surround you—the wind, the rain, the snap of twigs, the feel of loving company, and the music that connects you to all things.

It takes concentration to open up. It takes a certain amount of training. Awareness is not a gift that certain people have and others don't. It is an art that anyone can learn with practice and dedication. People often ask me how I can tell there are dogs on the road beyond my Pine Barrens camp. I can tell because I have spent hours and days and weeks and years listening to the shapes of the sound that emanate from different parts of the Pine Barrens. I have familiarized myself with its symphony. I know the melodies played by its instruments, and its notes are engraved on my heart.

Some of my students are perplexed and frustrated when I say this. But the fact is that we all receive the same stimuli from the world. The same sights, sounds, smells, tastes, and sensations that I pick up in the Pine Barrens are also imprinted on the minds of everyone else who goes there. The difference is that many people cannot bring them into the conscious mind. The walls of habit and logic are too thick. But fortunately these walls can be broken down.

There are so many levels of awareness. On a logical, scientific level you can look at a campfire and say, "This fire is burning because the fuel has reached the combustion point, and carbon and oxygen are combining to produce carbon dioxide, water, and heat." On a survival level you can say, "This fire is made with a certain type of wood, which gives a certain kind of light under certain conditions of dampness and temperature. It smokes buckskin a certain color, and it makes food taste a particular way." Then we can go to a more philosophical level and say, "This isn't fire at all. It's wood releasing years of stored sunshine." And on a spiritual level, you might say, "These flames remind me of the glow of the Great Spirit. They have come to warm my lodge and wash away all negativity and fear."

These are all valid ways of looking at fire, and there are many more. The important thing is that objectivity is not the only perspective. Don't be imprisoned by logic. Open up your intuition and let your gut feelings come through. When you do, you will begin to feel the spirit-that-flows-through-all-things. And that will be the beginning of even deeper explorations.

Many people cut themselves off from deeper levels of experience because they are afraid to be spontaneous and take chances. I'm not saying you have to put your life in danger. But if you're tired of living a mediocre life—if the ruts of habit are so deep you can't see over the sides—it's time for a little adventure. What that means depends on you. You might climb a tree, go down to a cold stream and go skinny dipping, jump into the muddy ooze of an inviting swamp, or just part the beautiful earth and smell its rich aroma. The point is, don't be satisfied just to sit and watch everything go by. Go out and live it!

Open up your senses, too. Most of the students who come to my Pine Barrens classes are aware of the wild dogs that live there, though the dogs rarely visit my camp. When I get complaints from people who think their senses are closed down, I sometimes invite them to take a hike to the dump wearing a pork-chop necklace. So far, no one has taken me up on the offer, though I can guarantee that, if they did, they would find their senses were very keen indeed.

Honing Your Awareness

Fear is a great teacher, but there are other ways of sharpening the senses. Some of these methods are discussed in the first part of *Tom Brown's Field Guide to Nature Observation and Tracking*. I would like to summarize and add to them here.

First of all, those who think they can simply go out into the woods and learn the arts of seeing, hearing, tracking or stalking without practice are sadly mistaken. Nature awareness requires the same kind of concentration as any other art. If you're teaching someone to play tennis, you don't just say, "Here's a tennis racket and here's a ball; now see if you can get the ball over the net." You show them the different strokes and have them practice each one repeatedly, emphasizing the proper form and footing. Then you begin putting the strokes together. When the strokes become second nature and the student can flow smoothly from one to another, you begin talking strategy and getting involved in the more creative aspects of the game.

Before you can master the game, you have to master the fundamentals. The same is true of nature awareness. A young Indian scout did not learn how to observe nature by walking a trail or sitting under a pine tree. He was taken aside by one of the tribal elders and pushed relentlessly to the limits of his ability. He learned the fundamentals thoroughly because he knew his life and the lives of many others would depend on them.

Opening the Senses. There are some simple exercises you can do to hone your senses, regardless of whether you're in the woods for fun or for a long-term survival stay. The first and most important exercise is to relax. Just lie or sit in a comfortable position in a place you enjoy. Close your eyes and let go. Then take a few deep breaths.

There are two ways to take a breath. One is like a bum with a bottle of Thunderbird, and the other is like a connoisseur of French wines. Be the connoisseur. Sit back and savor your breath. Don't try to smell anything. Just let the air flow in and out, allowing it to bring you news of whatever it contains.

Next, open up your ears and listen. Again, don't push yourself. Just be receptive to the whole symphony of sounds. Don't concentrate on one sound over another. Listen to everything. Get down below the baseline din and discover the instruments you may never have noticed before. Then savor the purity of the sound and hear how it all blends. Finally, notice the "width" of sounds. Feel the hollow places, the echoes, the quiet melodies, and the sharp, off-key notes and begin to draw a symphonic landscape.

Still with your eyes closed, feel with your whole body. Don't concentrate your sense of touch only in the fingers. Let your whole body come alive with feeling. Feel the damp breath of the wind against your cheek. Feel the pressure and texture of the clothes on your body and the ground beneath you. Map these sensations, too.

Next, open your eyes and look around. Do not label or judge anything. Look at the reflection of the fire and see it for the dance that it is. See the warm glow on your earthshelter or on the faces of your friends. Look at the color of the ground and the way the tracks are enlivened by the shadows. Look at the shapes of the trees and stars and the shadows of the night.

Finally, put it all together. Take a deep breath with your eyes open. Feel, smell, and taste the wind. See and hear the world around you with a new openness. As you do this, realize that you can become more open still, that there is no limit to how far your awareness can expand.

Have you ever really touched a leaf, a pine needle, or a slab of bark? Have you ever felt with your entire body what the day is telling you? Have you ever listened intently to the whole sphere of nature? Don't dwell only on the obvious. Don't be bound by time, space, or thought. Go deeper, always deeper. Let your own purpose be like that expressed by William Blake: "To see the world in a grain of sand, And heaven in a wild flower; Hold infinity in the palm of your hand, And eternity in an hour."

As you do these exercises, allow the spirit of the woods to flow through you. Imagine it as a soothing wind that permeates everything. Feel yourself resonating with it. That is what Stalking Wolf always did, and I believe it brought him closer to the truths it contained. Stalking

Wolf would sometimes stop beside a bush and look at a leaf as though he were talking to it. He would rub it between his fingers, feel the dust that had collected, and watch how the light played on it. He would notice the colors, textures, shadows, shapes, and geometric designs, and how it was attached to the plant. Sometimes he would take the time to smell it or taste it. I could see him reaching out to it with his heart, even though no words were spoken.

Concentric Rings

Another way of getting closer to nature is to become acutely aware of disturbances in the life flow. What happens when you throw a stone into a quiet pond? Initially there is the splash. Then a series of concentric waves begin emanating from the point of impact. They keep flowing outward, growing larger. On their journey they affect everything they pass. Insects are lifted up as though on ocean swells. Pond lilies ripple. Fish dart for cover. Sand grains shift on the shores. Then the waves begin flowing back to their point of origin. The impact of a

single stone causes thousands of disturbances, and each of these may cause thousands of others, until the surface of the pond finally smooths out and returns to normal.

The same thing happens in the woods. A fox walks into an open glen and it's noticed by little birds that flit and peep in the branches above it. Soon the jays are attracted and they begin screaming at the fox. A deer hears the jay's calls, jerks its head up, and perks its ears forward. Sensing this, a deer close to the fox does the same, and the fox stops in his tracks. The ripples return, and the fox is affected by its own concentric rings.

This kind of thing is going on everywhere, all the time. But the system is more complicated than it seems. Nature is not the flat surface of a pond. It is multidimensional. The disturbances that radiate are not just those of the physical senses, but vibrations of the spirit-that-moves-through-all-things. On the deepest level, all is one, and all is felt and known within the boundaries of everything else.

In terms of nature awareness, what this means is that you can learn to attune yourself to the symphony of the woods and begin translating the disturbances that are occurring at greater and greater distances. You can take in subtler and subtler nuances of meaning. It is important to first become familiar with the baseline, the pervasive melody that predominates at any given place and time. Once you are familiar with the baseline, you can easily recognize when a "stone" is dropped in the pond or when one of the instruments in the symphony hits an off-key note that causes confusion and discord.

Remember, though, learning concentric rings is not like sitting at home listening to a symphony with a set of earphones. To really appreciate a symphony, you have to be there—not just listening to the music, but watching the movement of the instruments, feeling the electricity in the air, even smelling the odors of the concert hall. The wilderness is just such a symphony. It is a multidimensional tapestry of concentric rings. Some are caused by animals, some by plants, some by wind and temperature, some by the changes of the seasons, and some by the march of time and geologic events. The more rings you can read, the more easily you can put together the puzzle of what is going on.

The first thing I do when I approach an unfamiliar area is sit down and wash all tension from my body and mind. A good meditation helps initially, though after a while this becomes a matter of habit. In about ten to fifteen minutes the concentric rings of disturbance I have caused by walking in have dissipated, and the symphony begins. After twenty minutes of concentrated effort, I know pretty much what the baseline is. Everything I sense afterwards will be measured against that baseline.

Most important, I consider the whole picture. I don't concentrate on sounds and sights to the exclusion of smells and sensations, and I don't focus on specific birds or animals because that is too narrow a scope. I try to imagine the land, air, and water around me as a unified sphere of influence, and I try to take in the whole picture without becoming distracted by the parts.

When you first do this exercise, you will probably be able to read only a few rings, so to speak. But eventually you'll become more sensitive and be able to read farther and farther out, until finally you will find yourself jolted by sensations within your body that do not even have names. You will begin to open up the sixth sense—the gut feeling that comes as a hunch or a flutter of the heart. Eventually you will begin to trust these impulses as deeply as those of the five physical senses.

When you sense a disturbance that you can't identify, go look and see what it is. That's how to learn. You have to take the initiative. I remember one class I had where every evening a yellow-billed cuckoo and two different species of tree frogs would begin chirping and croaking in the trees, and my students asked me over and over again, "Tom, what's that sound?" If I had a dollar for every time that question was asked, I would probably own the Pine Barrens. But what a worthless question! I would rather have died than ask Stalking Wolf a question like that. Nothing worthwhile comes easily. You can't learn nature awareness by sitting in an armchair. You have to earn it through effort and dedication.

Landscaping

One of the most powerful tools for nature awareness is habitually asking one simple question: "What is this landscape telling me?" This one question, followed by enough exploration and meditative thought, will give you the answer to almost anything you want to know about nature.

Another thing I do when I come into a new area is familiarize myself with the lay of the land. In my imagination I fly up and try to see it as through the eye of an eagle. Then I verbalize to myself, "I'm in the Pine Barrens, and I'm on a gradual slope that rises into the foothills of a low mountain range. There are a few grandfather pines here, but most of them are "teenagers," plus a few small ones. Mixed in with the pines are various species of scrub oak, which gradually give way to blueberries, briar, grackleweed, cedar, and sweet gum as I approach the swamp.

Then I begin to get more specific. I see that some parts of the area are thin and have been burned over, maybe five or six years ago. I can tell that in these areas the acorn crop is going to be good. It will get even better as I approach the swamp. The pines are also open, so I know I'm going to find deer, a good assortment of squirrels, rabbits, and mice, as well as foxes, coyotes, owls, and hawks. This gives me a feel for the area's history and the life it contains.

In the same way, I can also get a feel for the effects that humans have had on the area. I see a road and I ask, "What kind of machines were used to make this road? How long ago was it made? Why was it made? What does it tell me that it comes in here and dead-ends? How about trails? Where do they come from? Where do they go? Why are they worn in this particular way?"

With my overview and a little thought, it may occur to me that the road was made by a bulldozer. From the size of the bushes and trees just beside the road, I can tell that it is about five or six years old. Since it dead-ends in the cedar swamp, where a lot of trees have been taken out, I conclude that it's mainly used as a logging road. The deeper ditches are firebreaks, and the trails are made by loggers and campers who have left quite a few other signs.

Once I get an overview of the area and its history, I begin to get more specific. Now I might look at a single tree and ask myself, "What is this tree telling me?" There is a message in every branch. There is a reason for every crook or discoloration of leaves. I see a branch that has been broken off, and I know it was broken off live because of the sap that has collected. I ask what instrument was used, and the shape of the break tells me that it was a sharp blade such as a machete, and that it took two whacks to cut it off.

Now I look at the leaves and see that some are turning color while others are not. After some quiet observation I discover that the leaves in direct sunlight are still green, while those that are shaded have begun to turn color. From the leaves of one tree I can learn something that is universal about the natural world.

I can also learn very particular things that are just as fascinating. One fall when I was a boy, I was perplexed by a pine-tree branch whose bark had been worn off in one place. I was mystified about what might have caused it. I kept watching the branch at different times of the day to see if anything came by. "Could it be a squirrel?" I asked. No, there were no signs of food having been eaten and no good reason for a squirrel to return to that spot as many times as it would have had to in order to scrape off that much bark.

I kept coming back to the spot and watching, and nothing came. Then I thought, "Could it be a bird?" Yes, I could see the clawmarks, and I did not think they were those of any mammal. But in the fall landscape I could see no reason why a bird might perch and fly off and return so many times from that particular spot.

It took until spring to answer my question, but I did not give up. With the passing of winter and the melting snow, a small swampy area with a spring was created around the pine tree. As it filled with water, it became a breeding ground for mosquitoes and other insects. At different times during the spring and summer months, that branch became the favorite perching spot of two different flycatchers: a phoebe and an eastern kingbird. Both like to return to the same perch again and again after hawking small insects on the wing.

It took me seven months to learn the answer to my question. But I finally found the answer because it burned inside me. My inner yearning led me not only to the specific answer I wanted, but gave me answers to a wealth of questions I had not even intended to ask along the way. You never know where one question will lead. It is not so much the answers as the quest that is important. If you keep asking and seeking, you will be led down paths of adventure you never imagined. It's all learning, and it's a joy every step of the way.

Becoming Invisible

All things in nature know instinctively how to blend and flow. Animals and insects wear colors that let them disappear against a backdrop of leaves and twigs. Trees and grasses know how to let the wind sift through their blades and branches. Even rocks allow themselves to be molded by wind and water to conform to their places in the environment. All natural entities move and grow with the seasons and in relation to one another in such a way that there is no disharmony. It was from these natural entities that our ancestors learned how to blend and flow with the woods.

On a physical level, the art of movement and camouflage is fairly simple. For those who are interested, I have outlined the basics of these arts in *Tom Brown's Field Guide to Nature Observation and Tracking*. There is actually little more that needs to be said about them because, once you are living in a survival situation, the land will teach you all you need to know. But I would like to summarize here and add a few more details, especially with regard to camouflage and the mental aspects of blending and flowing.

Most important, you should move very slowly and quietly, lifting the feet high and setting them down so that you feel the earth beneath your feet. In a way, you begin to "see" with your feet, knowing instinctively how to avoid a rock or a twig that might break, and keeping your eyes on the horizon to take in the maximum amount of landscape.

As you move, be intensely aware of everything that is going on around you. Reach out with your senses, putting them all together. The more this becomes ingrained, the more connected you will feel. As you move through the woods, you will begin to sense with your whole body what is going on around you. Eventually you may be able to stalk through the woods blindfolded, avoiding obstacles almost as if you had seen them. This is nothing magic or supernatural. It is simply a reactivation of the ancient instincts. It is a reconnection with the earth and the spirit-that-moves-through-all-things.

The physical art of camouflage is broken into several parts. First comes treating the major scent areas of the body (forehead, chin, armpits, and chest) with a natural plant that masks the human smell. Some of the best are wintergreen, mint, and crushed cedar berries, though any plant that occurs in the area will help. It also helps to chew on a wintergreen berry or some other pungent plant to get rid of "bad breath."

Next comes shadowing, which is basically dulling the body with a thin layer of earth, mud, or ash. Cover almost the entire body, particularly the high spots that reflect the sunlight. Then comes dappling, which is using a darker material such as mud or charcoal to dull and break up your outline. It's best to apply this to the high spots on forehead, nose, chin, beneath the eyes, the other prominent parts of the body so that your outline begins to look like it's lost in shadows.

Finally comes fuzzing, in which you get down in the dirt and roll around, picking up pine needles, leaves, moss, and debris. In this stage, you can tousle your hair and decorate it with a few leaves or strands of moss, but don't overdo it. A little goes a long way.

You have to blend with the area. Always use materials that are taken from near the area you plan to use, never from some foreign spot. If you use skunk cabbage to camouflage your scent in the upland, or oak leaves to disappear in a swamp, you're not going to fool anybody. Use

**A dappled,
irregular pattern makes
the best camouflage.**

the materials that are already there.

It doesn't take a lot of leaves or dirt to make the body disappear. One fall I had ten students disappear, all within arm's length of the trail, under nothing but a thin dappling of leaves, twigs, and sand. I had to point them out one by one to a camera crew that was filming the whole process, and I even missed one myself. Their secret was not in burying themselves, but in becoming part of the landscape. And to do this, they really had to play the part. They had to work on their inner camouflage.

Whether tracking, stalking, or lying in wait, as long as you feel separate from the landscape, you will not blend with it. Because of your inner feeling of separateness, you will do something to make yourself visible. On the other hand, if you have a sense of belonging, you will tap into the flow of the area and automatically do what needs to be done to disappear.

In other words, camouflage takes not only skill but an open heart. Those who are able to disappear do not put themselves above the environment. On some level, they communicate with the soil and plants and rocks. They ask how to become part of the natural pattern. Those who ask in a spirit of humility and sincerity are given the greatest rewards.

When you are settled, relax completely and take a deep breath. Let it out as though somebody had just pulled a plug inside. As you let go, allow your inner self to blend with the area. If you're lying beside a stump, become the stump. If you're tangled up in some branches, really feel your arms and legs becoming branches.

There is a communication beyond words in the natural world. If you're excited or upset or throwing off vibrations that are not in tune with your surroundings, the animals are going to feel it and avoid that area. So don't put any kinks in the life flow. Blend with it, and you will discover delights and surprises you never expected.

A Parting Word

The depth of awareness and connection you reach with the wilderness world depends on you. There is no limit to the levels you can experience. No matter where you are, you can always go deeper. But the deeper levels do not come without effort. Many of my students seem to want a magic pill that will automatically transport them to a kind of nature nirvana without their having to put out any effort. Other people seem to feel that the wisdom of the universe can be absorbed through the pages of a book. It doesn't work that way. There is no magic pill. The magic is created inside, through the power of your own will. And the essence of that magic is fire. It's a burning desire to push deeper—to know and feel and experience more.

Most people, when they see a deer run off a trail, will watch until it disappears and then go on their way. When I was learning to look at nature I would watch and listen long after the deer had disappeared. I would stand for an hour, two hours—however long it took—until I could pick out an ear, a foot, or some fleeting movement that showed me where the deer was. Then I would watch the animal again until it disappeared, and repeat the whole process. During this game the deer would gradually lead me deeper into the forest, but it would also reveal more of its secrets along the way. The essence of the deer would become more visible to me, until I could begin to see it not only with my eyes but through my inner vision.

On one level, nature awareness is a skill. It is knowing the names of plants and animals, becoming familiar with their habits, knowing how to read the landscape, and so on. But on a much more powerful level, it is the ability to open up and communicate with your heart. Once you get a taste of this, there is nothing that can hold you back. The faint flicker of desire that was kindled deep within may turn into a flame, and that flame may grow into a warm fire. Finally it may turn into a roaring inferno that cannot be put out by the strongest winds and storms.

11
CIRCLE OF THE SEASONS

One of the greatest and most rewarding challenges a survivalist can face is to attempt to live for an entire year off the fruits of the land, without any of the trappings of civilization. There is no way I know of that brings a person more in tune with the true rhythm of the seasons and the cycles of the earth. The very commitment strengthens one's resolve, and the experience itself brings one into such a close harmony with all living things that I cannot begin to describe the gifts that are received from it.

Perhaps you are a person who has yearned to feel what it is like to truly live off the land—or, as I prefer to put it, to live with the earth—for an extended period of time. Whatever the time period, you will gain greater wisdom and insight into your own life and the life of Earth Mother. Even a short-term survival situation can be a rewarding challenge if you go into it with the attitude that it may be an opportunity for you to discover some things about yourself and your place on the planet.

With this in mind, I would like to spend a few pages discussing the discipline necessary for survival living, and in particular for extended earth living. I will begin by describing the sequence of activities I would go through to obtain my necessities during the first three days of a primitive survival situation, then go on to describe the kinds of activities that would be necessary during each of the four seasons. This way, you will have a general idea of what the survival priorities are, regardless of how long your wilderness stay might be.

(*Note*: I do not recommend trying an extended survival stay until you have become proficient at basic survival skills. For detailed information on basic wilderness living, see *Tom Brown's Field Guide to Wilderness Survival*.)

The First Three Days

There is no telling when you will find yourself in a survival situation. However, for purposes of simplicity, I'm going to begin this description as though I were going into the woods on the morning of a bright spring day. During the first day, the most important priority is shelter. But while I'm gathering shelter material, I also think about other needs such as fire and food. In other words, I try to tend to several

207

things at once. That way I don't waste precious time and energy, especially in bad weather. As I'm collecting poles, brush, and insulation for my leaf hut, I also have my eye out for wild edible plants and materials for a bow-drill fire. These I pick up along with the materials for the shelter, stuffing the tinder and edibles into my pockets as I find them. Unless I'm in a dire survival situation, I take these materials only from areas where they're abundant. In other words, I don't strip all the bark off the only cedar tree around, and I don't pick the last of the wintergreen that's growing in a patch all by itself. Whenever possible, I maintain the attitude of a gardener, trying to thin out plants where they are already growing in profusion. I also remember that the plants are my brothers, and I silently send them thanks for their contribution to my well-being.

Likewise, during and after the construction of the shelter, I keep an eye out for small, wild animals such as squirrels and rabbits and curious birds that may provide a relatively easy meal either through hunting or trapping. From the first hour, I am never without a primitive throwing stick or a few rocks that will fit easily into my hand or pocket. I also scout out my trapping areas from a distance but do not disturb them unnecessarily by going into them. I try to stay intensely aware of everything that is going on around me. With basic tracking and nature awareness skills, I familiarize myself with the sights, sounds, smells, and feel of the area. I ask myself what animals it might support and where they are most likely to be found.

By the end of the first day, I have usually built a shelter, made a fire, killed and skinned at least one small animal, begun to coal-burn some eating utensils, woven a crude burden basket, and made approximately ten traps—mostly light, easily-portable snares. In addition, I have scouted out the best trapping and hunting areas in preparation for the next day's activities.

On the morning of the second day—ideally, after a long, sound sleep in a well-sheltered bed of dry leaves and mosses—I awake before sunup and go out to set my traps. Usually this takes a good, long while because traps are most effective if they're set some distance apart. This is both because of the size of animal territories and because of the fear and confusion that's caused when one of the traps is sprung. Once a trigger is tripped, all the animals in that area go on the alert. Never assume that animals don't talk to each other. There is a complicated and intricate communication among animals that tells them all they need to know.

The setting of traps and the traveling associated with it may take most of the day. Partly for this reason, I put the trap materials in a large basket or container. While I'm setting traps, my travels are taking me through areas that are rich in wild edibles. This is only natural, as most of the small rodents I want to catch are attracted to the best feeding and watering areas. There are almost invariably transition areas such as the fringes of forests, meadows, and streams. As the edibles appear, I pick them and put them into the container for the return trip.

When I return to camp that night, I stoke the fire, hang the wild edibles up in the shelter, and prepare a meal. That evening, after a hearty meal of rabbit stew with cattail shoots and other delicacies, I finish up the wooden bowls and eating utensils and work on some more baskets. I may also begin scraping and tanning the hides of a few small animals. These I will later make into clothing articles—a loincloth, a hat, gloves, or whatever I need most.

On the morning of the third day, perhaps with a strong burden basket on my shoulders, I'll go out and see what I've caught in my traps. On my way, I pick up whatever animals have been caught and reset the traps. Again, I stay on the lookout for wild edibles, and also for the easy shot with a rock or throwing stick. I also begin looking for materials I can use to make crude weapons such as firehardened spears and short-term compound bows and arrows (described in *Tom Brown's Field Guide to Wilderness Survival*).

On the afternoon of the third day, I spend most of my time skinning and cleaning the animals, slicing the meat and putting it on drying racks around the fire. I also get the hides scraped and ready for tanning. That night I go to bed fat and happy.

From the third day on, survival becomes more and more enjoyable because by this time I have taken care of most of my needs and can now concentrate on my wants. It may take longer to reach this point, depending on your survival abilities, but with variations for weather and season, the sequence of activities is basically the same.

By the beginning of the fourth day, I feel such a familiarity with the area and what it has to offer that I've begun to feel a part of it. Even the animals now have accepted and begun to adjust to my presence. On this day I build an extension to my leaf hut, which is a workroom and storage room. This I will use to make tools and furniture and to store meat and hides that have been scraped. I'll also begin making myself a kind of pelt skirt, as well as my first survival bow and arrow. Finally, I'll begin to breathe a little easier. I'll look around and enjoy the clouds and sunsets more. I'll take more time to feel the rhythm of the earth, and to give thanks to the Creator for smiling down on me.

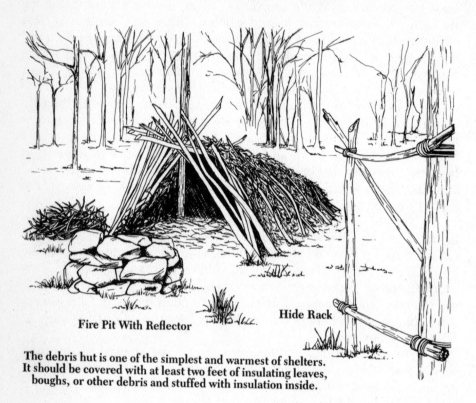

Fire Pit With Reflector   Hide Rack

The debris hut is one of the simplest and warmest of shelters. It should be covered with at least two feet of insulating leaves, boughs, or other debris and stuffed with insulation inside.

Long-Term Earth Living

Now comes the turning point. By the beginning of the fourth day, I usually have a substantial shelter and enough meat and plants to keep me fed for at least a week. I also have several bowls and baskets. My shelter has grown, and so has the amount of time I have on my hands. In general, there is a lot of leisure time in a survival situation. But in a way this is an illusion, because nature is always changing. Every season has its demands, and to survive and thrive near the heart of Earth Mother you have to take good advantage of what each season has to offer.

The real turning point for me—the point that marks the difference between survival living and earth living—is the hunt for my first deer or other large animal. Once I get my first deer, I'm set for at least a month. The deer provides everything I need for tools, clothing, and food. It is a tremendous source of security. After this I am ready to build a long-term shelter and to settle down into the more leisurely aspects of earth living. I still continue my trapping and hunting forays on good weather days, but now I also begin to spend more time at home, working on shelter improvements such as grass mats and shingling, and on other crafts that can be done around the fire. I retain the leaf hut as a storage and emergency shelter, but now most of my energy goes into careful craftsmanship in shelter and tools.

During these more leisurely days, unless I'm stuck in the middle of winter, I never try to rush. Generally, I go hunting in the morning, take a midday siesta the way the animals do, and spend the afternoon and evening on shelter building and crafts. As much as possible I gear myself to the natural rhythms around me. I also take my big meal in the middle of the afternoon, because I can see better to cook. At night I usually munch on leftovers and spend a lot of time working around the fire on reed mats and other things that don't require a lot of visibility. When I reach this point, I know I've made the transition from survival to full earth living.

This transition may sound easy, but it isn't. The first three days in full survival are very intense. They demand a positive attitude and well-honed skills. If you have not reached this point and you would like to try survival living, I would strongly suggest that you carefully master one skill at a time before you try to put it all together. Only when you have mastered the basic skills will you be ready to begin dealing with seasonal changes.

Regardless of the season, though, try to make the area your home. Treat it as you would your own property. Pay it some respect and try to beautify it. If you want to camouflage your shelter area, take care in the way you enter and exit the area, making thin trails and not trampling the brush. Also, don't harvest too many plants and animals in any one area, and don't harvest at all in the immediate vicinity of your shelter. This area should be reserved as a sanctuary—not just for aesthetic reasons, but as an emergency food reserve. It's like a storage bin or refrigerator. If you keep it well stocked, it will always be there when you need it.

Spring. Spring is an excellent time to begin an experiment in earth living. It is a time when green shoots are sprouting, when animals awaken and begin to forage, and when birds return to their nesting areas. It is a time of new life for the earth and all its creatures. Most of all, it's a release from the cold of winter and the return to the warm embrace of the sun.

During the spring your main concerns are gathering lots of wild edibles and materials for tools. It is a time for repairing bows and arrows, adding on to worn-out shelters, airing out and storing mats and hides, and repairing things that are frayed or broken. It's also a time to get outside more and begin to move as the animals move. It's a time for exercising muscles and streamlining the body. It's also a time to gather new greens, enjoy some fresh meats, and begin to sow some of the seeds that were collected from the previous fall season.

Summer. During the summer it's usually very easy to meet your survival needs. It's relatively relaxed—a time to kick back and enjoy the fruits of your labors. But it is also a time of preparation. Most winter preparations are made in the fall, but you can't wait until the last minute for all of them. Drifting into summer, then, you must begin the long but gradual process of putting things by for the colder months. You'll do some basic repairs on your shelter, strengthen poles and lashings, add insulation and roofing, or even move camp and build a new shelter in a better location. You'll also take care of your water source to make sure you'll have plenty of liquid in the midst of winter.

Summer is also a great time to dry and store plants for teas and spices, to smoke meat, collect seeds and basket and matting materials, and to grind grains. It's also the ideal time for tanning hides and making earthen pots, since you can take greatest advantage of the warmth of the sun. By the end of summer, you will have also gathered enough materials to begin making such things as long-term bows and arrows and other finely crafted tools.

Fall. With the onset of the fall season, there's an intense frenzy of activity. You can see it in the animals, and you can feel it stirring in yourself. It is the final time of preparation for the cold months ahead. Fall requires the greatest discipline of all. You don't dare slack off, especially on good days. If you do, you're likely to have a miserable winter.

For this reason, as you get closer to fall, you'll want to make more intense repairs on your shelter, concentrating on the insulation. You'll also want to stack plenty of brush on the windward side of the shelter, and stack up plenty of extra firewood in a protected and convenient spot where it can season well.

Fall is also the time when all animals must somehow put away enough food to last them through the winter. For this reason, your hunting and food gathering will get more intense. A large part of your efforts will be taken up in the harvesting of roots and tubers as the plants begin to die off. You'll also hunt a great deal and begin stockpiling meats and hides and hanging them up to smoke. You'll start to collect acorns and other nuts. You'll store lots of grains and dry and smoke lots of meats in an effort to get a good cache set aside for the winter. You'll also collect lots of cordage materials from dead plants and the inner barks of trees.

Winter. Winter is the time when the earth shuts down and shrouds itself in protective snow. It is a time when plants stop growing and store their energies in roots and tubers. For this reason it's an excellent time to gather roots. But winter is also a time of introspection and indoor activity, when animals hole up in protective burrows and nests. It's a great time for home crafts such as the making or repairing of tools and clothing. Depending on your needs and inclinations, this may include anything from quillwork, beadwork, and basketry to pounding sinew and making rawhide implements and containers. There are endless opportunities for useful winter occupation. Hides can be freshened up and scraped. Lashings, shingling, and mats can be strengthened and repaired. Arrowheads can be knapped. Old arrows that show signs of fraying can be reshaved and straightened. And you can wrap cordage until your hands hurt.

February, when the sap is down, is an excellent time for harvesting bowstaves and arrow shafts. Generally you should hunt in the winter only when conditions are ideal—neither too wet nor too cold. Most of the time you should be living off what you have stored during the previous summer and fall.

The most important part of long-term earth living is planning ahead far enough so you'll be ready for the next season. It's knowing in advance just what demands the seasons will make on you and how to cope with them. This, of course, varies with the area. But, in general, you always have to prepare for the season ahead. If you do this, you will soon be traveling comfortably through the seasons just as the animals do. You'll not only be surviving but thriving, and truly enjoying your part in the great drama of nature.

12
CARETAKER

A caretaker does not own the environment in
which he lives. Strictly speaking, he does
not own anything, not even his body or mind or life.
Marian Mountain, The Zen Environment

There is an attitude of concern that goes along with earth living. It is not just making sure we don't get sick from polluted water, or that we consider the needs of somebody who might be camped farther downstream. It goes deeper than that. It is an attitude of being connected to all things and caring for them—like a caretaker or a master gardener.

A master gardener not only knows how to weed and mow and plant; he or she knows what each plant needs, where it grows best, and how it can contribute to the overall beauty of the garden. A good caretaker not only has responsibility for preventing damage and keeping the property clean, but also for protecting and maintaining it so that others can enjoy it. A good caretaker is sensitive to the needs of the whole environment.

We do not own the earth; we are its caretakers. We have a responsibility toward it. There was a time when we were so few that the earth could absorb and correct almost any mistake we made. But no more. Our power and influence is too great now. If we refuse to take care of the earth, or do a shoddy job of it, the effects of our mistakes will eventually come back to haunt us.

We already know this. Mercury poured into the oceans comes back to poison our fish. The smoke of industry comes back to poison our lakes. Chemical dumps that seem convenient one day turn into deadly nightmares the next. Why do these things happen? Because we don't always feel our connection to the earth. We don't always take care.

When I was eleven years old, I had been under Stalking Wolf's wing for more than four years, and I had begun to feel some of the earth's pain as my own. Stalking Wolf was always talking about how Earth Mother was sore from all man's chopping and digging. I had seen strip mines and clear-cuts, and it got so I couldn't see a beer bottle on the landscape without getting angry. Rick and I avoided roads and trails, partly because we learned more by walking through the woods, but also because we felt that they defiled the ground in some way.

Rick and I were extremists then, but our feelings served their purpose. Hating roads and trails for a while helped us to see more clearly what was sacrificed for their convenience, and in the end it helped us to appreciate them more. Our extremism also helped us to realize our own effect on the environment and our responsibility toward it.

When I was twelve, I decided that the earth was very sick, and getting sicker (a conviction I feel just as strongly today). Even around my own little town of Toms River, wild places were disappearing fast and being replaced by highways and parking lots. In many places, it seemed that the earth had been ruthlessly torn up without any thought for the plants and animals that lived there. It seemed like the only consideration was profit and what could be gained for today. But it got me to thinking: When the rivers became sewers, where would we get our water? When the sky turned black, where would we get our air? When the topsoil was gone, where would we get our food? I felt as though I belonged to a generation of people who were killing their grandchildren to feed their children.

With my child's understanding of things, I turned my anger inside-out and decided to do something to help heal the wounds of Earth Mother. I told Stalking Wolf I wanted to be an earth doctor. He nodded his approval, but as always he left it up to me to decide just what that meant. I started by picking up garbage and beer bottles and taking them to the dump. But my efforts were short-lived, because wherever I picked up a beer bottle one day, two of them seemed to show up the next day.

In my doctoring I learned more by watching how the earth healed itself than I did trying to do it on my own. I noticed that in places that had been scarred by bulldozers, certain plants like fireweed and thistle would pop up. Then came grasses, which would attract mice and voles. Before long the grasses would give way to bushes and trees, and other birds and animals would come in. In other words, I found that the earth, if given a chance, would eventually heal itself.

With this realization, I thought there might be some way I could cooperate with it. So I started carrying seed pods with me. I stuffed my pockets with amaranth, burdock, mullein, sunflowers, and other seeds that grow in waste places. I called these "scab plants," and I spread them over scarred landscapes wherever I went. Then I watched what happened. I saw which ones grew and which ones didn't, and I saw what effect they had on the other plants and animals that came in.

I learned a lot by playing earth doctor. But the most important thing I learned is that it was useless for me to do it alone. I realized before long that I could never heal the earth by myself. I would have to get help. So I got Rick to help me. But Rick and I together couldn't do it, either. We couldn't even make a dent in the Pine Barrens, let alone Toms River or the rest of New Jersey.

Finally I went to Stalking Wolf and asked him what I should do, and he said that all the scab plants in the world wouldn't help the earth unless people got their feet back to the soil. He said it was only people's hearts that could cure the ills of Earth Mother. Loving and caring was the only answer. He told me that some day I would go on to teach the old ways, and that this might open some people's hearts and help the earth in ways that garbage collecting and scab plants never could.

Since then I have realized that it is more important to cultivate people than to cultivate plants. It is better to lead people back to nature and let them feel their own roots than to pick up beer bottles and trash, because with a new awareness fewer bottles and trash got left in the first place. This goes for larger decisions, too. Once people begin to realize their earth connections, they begin to act in very different ways. They begin to feel responsible not only for their "own" little piece of property, but for all the property attached to it. They realize that they cannot abuse the land anywhere, because it belongs to them and they belong to it. All the earth is their own backyard.

One of the reasons I stress living simply in a small area is that it is easier to see how even the simplest decisions affect the environment. We can't walk to our shelter without starting to make a trail. We can't sit on the ground without mashing some grass or killing some insects. We can hardly do anything without affecting the landscape or the living things around us in some way.

It's very difficult to see the long-term effects of major decisions. It is even more difficult to see them if we don't understand our real needs. This is why survival skills and earth living are so important. They get us back to basics. They help us realize the true value of things like air and water and fire. Instead of building a big house on a big plot of ground, we can build a small shelter on a little piece of ground. We can spend a lot of time on that ground, giving and taking. There, we can see the effect of every decision we make, and we can begin to develop a refined sense of what is important.

With earth living, we learn to take more care. We first learn to take care of ourselves and our own needs. Then we learn to take care of our earthshelter and everything around it. Finally, as caretakers in the highest sense, we learn to give thanks for the many gifts of the earth, and to feel a connection even in those things that seem farthest removed from it. We also learn that we can contribute to the beauty and health of the earth—not just for our own sakes, but for the sake of everything that lives.

Some of the best examples of caretakers are the Hopi Indians of the Southwest. They have lived in the same area for centuries, taking only what they needed, living in an intricate balance with the land. Another example are the Amish people, who have some of the most beautifully enriched soils in the world. In an almost magical way, they seem to be able to take barren, rocky soil and make it into a virtual paradise. Experience has given them the knowledge, and love of the land has given them a mastery of the art form. To live well with the earth takes both.

Our ancient ancestors knew through instinct, intuition, and generations of trial and error just what effects their living would have on the land. When they made mistakes, they could always move on. There are too many of us to move on now. But we can learn to live well where we are. We can reawaken the love of the earth that makes living an art form, wherever we might be. With modern technology, we can also expand on the science of survival in ways that our ancestors never could. Practice in living with the earth can teach us not only to be more deliberate about what we take, but also how to give back in such a way that we end up giving more than we get. In the end, that's what caretaking and living with the earth are all about.

APPENDIX
UTILITARIAN PLANTS

Following is a descriptive list of utilitarian plants that can be used with the skills presented in this book. For further information on these plants, I recommend *A Field Guide to Trees and Shrubs*, by George A. Petrides and *A Field Guide to Wildflowers of Northeastern and North Central North America*, by Roger Tory Peterson and Margaret McKenny. Both books are published by Houghton Mifflin. For descriptions and information on more than a hundred of the most common wild edible plants found throughout North America, see *Tom Brown's Field Guide to Wilderness Survival*.

ALDER
Alnus spp.
Basket splints

ASPEN
Populus spp.
Cordage from inner bark

BASSWOOD
Tilia americana
Bowstaves

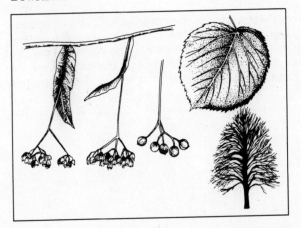

BIRCH
Betula papyrifera
Baskets from twigs and bark

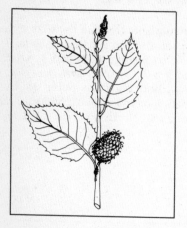

Description (for white birch): A medium-sized tree up to 80 feet, found across northern U.S. Characterized by white bark lined with faint horizontal stripes, with blackened areas around branches. Usually found in damp woodlands in close proximity to water. Leaves heart-shaped with fine serrations, from 1 to 4 inches long. Twigs when broken may have odor of wintergreen and bark can be separated into paperlike sheets. Flowers appear in spring, contained in catkins. In late summer these become long clusters of tiny, dried fruits.

Uses: The white birch makes some of the finest baskets and containers. The bark can be peeled and easily bent and sewn into containers of all sorts— even canoes. There are not many large birch trees left in the United States, but I find that the bark of fallen dead trees is just as good as that taken from live ones. Seams on containers can be made watertight by sealing them with pine pitch.

BLACK ASH
Fraxinus nigra
Bowstaves, cordage, basket splints

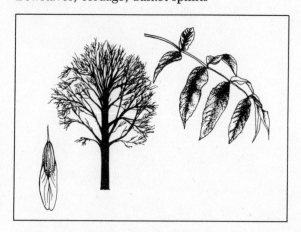

Description: Tree ranging from 50 to 80 feet tall. Often found in damp areas such as swamplands and floodplains. Leaves opposite and compound, 12 to 15 inches long, with 7 to 11 toothed leaflets and no stalks. Flowers appear in late spring. Fruits are rounded at both ends and ripen in summer.

Uses: The black ash is a multipurpose wood. It is excellent for fine bowmaking, being somewhat stronger than the hickories but just as easy to read, work, and bend. This makes it a particularly good wood to use for your first bow and arrow. The ash bow is best made as a long bow with a slight curve to the ears. This gives it a better snap when fired. Ash can also be used to make cross-country skis, snowshoes, paddles, and many other camp necessities. The inner bark of the ash can be collected from a newly dead tree or sapling, producing a good, strong cordage for binding and sewing. Ash logs can be soaked for several days and pounded into durable, long-lasting splints. By far my favorite baskets are made from ash—especially the larger pack and burden baskets.

BLACK WALNUT
Juglans nigra
Bowstaves

BLUEBERRY
Vaccinium corymbosum
Arrows, baskets

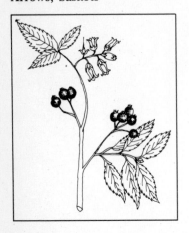

Description: A tall shrub growing in sandy, acidic soils. Leaves alternate, simple, elliptical, and usually without teeth. Flowers in late spring, and in midsummer bears the familiar blue to blue-black fruit with whitish, powdery glaze.

Uses: In many survival situations and in long-term shelters in the Pine Barrens of New Jersey, I have come to know the blueberry as a welcome friend for food, medicine, and arrowshafts, among other uses. The arrow shafts are a little heavy but very difficult to break, which makes them excellent for close-range hunting. Blueberry shoots and saplings also make quite strong and serviceable sapling baskets. One blueberry basket that I made from split saplings eighteen years ago is still in wonderful shape.

BOX ELDER
Acer negundo
Cordage from inner bark

BULRUSH
Scirpus spp.
Cordage and baskets from leaves,
basket splints from stalks

CATTAIL
Typha spp.
Cordage, baskets, mats

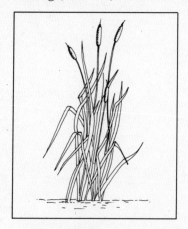

Description: Tall plants found in marshes and wetlands throughout the United States and Canada. Grow to 9 feet in dense colonies. Leaves lance-shaped, often as tall as the plant itself. Brown, sausage-shaped heads found atop the single long stalk from late summer through the following spring. In early summer a tube-like flower head and pollen spike can be found, located above the flower head.

Uses: Very similar to those of the common reed, *Phragmites spp.* (see page 240). Leaves of the cattail can be collected green and woven into useful mats for flooring or walls. They can also be used in the making of coiled or woven baskets. The stems can be bundled and tied to form thick sleeping mats, to line the walls of a shelter, or for thatching material.

CEDAR
Chamaecyparis spp.
Cordage, baskets, mats, clothing

Description: Atlantic white cedar is a medium-sized tree, growing to about 50 feet in swamps and heavy lime soils east of the Mississippi. Leaves are scalelike, arranged in four rows around the twigs, but appearing flattened from the sides. The fruit is bell-shaped and bluish.

Uses: Most species of cedar make serviceable arrows if from a good, straight sapling. The wood is quite easy to work, shape, and bend. I have made good shafts by carving down a splint broken from the main trunks of dead trees. Cedar also makes good splints for baskets. Splints are made by soaking the log for several weeks, then pounding the ends with a hardwood mallet, splintering the cedar at the growth rings into thin strips. These can be soaked again or steamed to make more flexible material for weaving baskets.

The twigs and small branches can be used to make sapling baskets. The inner bark of the cedar has many uses, including cordage, basket material, mats, blankets, and clothing. The best way to collect and prepare the cordage is to skin the bark off a dead tree and soak it for several hours. The inner bark can then be easily removed and laid out to dry. The fibers are best worked when lightly damp and can produce rope, cordage, or bundles of any thickness for a variety of uses. Cedar makes only a medium-strength cordage, but cedar mats, blankets, and clothing are soft, resilient, and durable.

CHERRY
Prunus spp.
Arrows, cordage from inner bark,
splint baskets from twigs and small branches

COTTONWOOD
Populus spp.
Cordage from inner bark,
splint baskets from branches

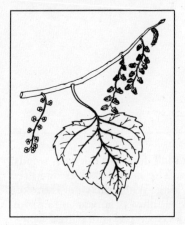

DOGBANE
Apocynum spp.
Cordage

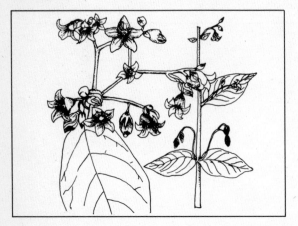

Description: A common plant, up to 4 feet tall, found in thickets, woods, and waste places throughout the United States. The hearty stalk is deep red and bears paired, opposite leaves, 2 to 6 inches long. When the stem is broken it exudes a poisonous, milky juice. The bell-shaped flowers appear in midsummer, white to pale pink, sometimes striped inside. Seed pods are paired, lance-. shaped, up to 8 inches.

Uses: Dogbane produces one of the strongest, most durable cordages, and it can be used for the most demanding jobs—even for strings on heavy bows.

DOGWOOD
Cornus spp.
Bowstaves

ELM
Ulmus spp.
Bowstaves

EVENING PRIMROSE
Oenothera biennis
Cordage

Description: A tall plant reaching to 5 feet. Found in dry soils and open areas throughout the United States. Flowers bloom in late summer, yellow with four broad petals, about 1 inch across. Below the flower petals are 4 reflexed sepals. There is also a crown-shaped stigma in the flower center. Leaves entire, lance-shaped, up to 3 inches long and densely covering the stem.

Uses: Makes a good cordage.

FIREWEED
Epilobium angustifolium
Cordage

Description: A tall plant growing to 7 feet, found in waste areas, clearings, and transition areas throughout the United States and Canada. Gets its name from frequently being among the first plants to appear in burned areas after a fire. Flowers appear in late summer, pink with 4 round petals. Leaves narrow and lance-shaped, alternate and entire, up to 4 inches long. After the flowers fall away, the seed pods appear at the top of the stalk and angle upwards.

Uses: Fireweed makes a good cordage.

GRASSES
Graminiae spp.
Coil baskets

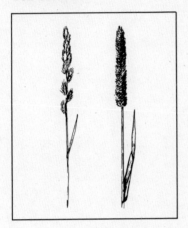

HICKORY
Carya spp.
Bowstaves, cordage, baskets

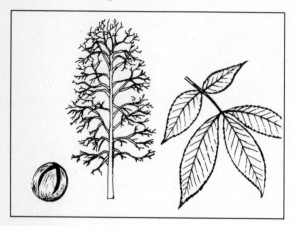

Description: Deciduous tree, often reaching a height of 100 feet. Found in Eastern United States from Maine to Georgia. Leaves alternately compound, with 5 to 17 leaflets. At summer's end the large, greenish nuts can be seen hanging in pairs. Husks of nuts break into four separate parts.

Uses: Though not as strong or resilient as Osage orange or yew, the hickory makes a fine, straight bow. It's an especially good wood for beginning bowmakers because the grain is soft and easy to work. Another advantage is that it need be seasoned for only six months prior to use. Hickories are best treated with many applications of hot grease to insure that the wood keeps its suppleness and spring. The inner bark can also be used to make cordage, and the branches and twigs make good, thick splints for baskets.

HONEYSUCKLE
Lonicera spp.
Baskets

Description: A creeping or climbing vine with hollow pith, found in thickets and forest edges. Leaves opposite, simple, and untoothed. In some species the leaves may be joined at the base, encircling the stem. The flowers range from yellowish to white and in most species have 1 lower and 4 upper lobes at the opening of the tube-shaped flower.

Uses: Honeysuckle is best for weaving baskets, but the newly-dead vines can be soaked for several hours to bring back the suppleness. Honeysuckle baskets are best for light loads or interwoven with stronger fibers to make baskets for heavier loads.

IRONWOOD
Carpinus caroliniana
Bowstaves

JUNEBERRIES
Amelanchier spp.
Arrows

Description: Juneberries (serviceberries) are usually shrubs but may grow to the size of a small tree. Their leaves are alternate, simple, toothed, and many times blunt-tipped. Buds are pink to reddish, depending on the species, and the bark is quite dark.

Uses: Most serviceberries make quite good arrow wood. They are easily worked, bent, and shaped. They hold their shape remarkably well, even when stored poorly for long periods of time. I also find them quite good when fashioned into quick survival arrows, since they need little work other than removing the bark and straightening. Juneberry arrow shafts can be completed in just a few minutes, but for long term use they should be made in the usual way.

JUNIPER
Junipereus communis
Bows, cordage, baskets from rootlets

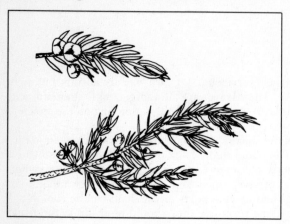

Description: Shrub reaching 5 feet in height. Leaves are whitish above, scalelike, and 3-sided, occuring in whorls of three. Twigs are also 3-sided. Fruits usually hard and rounded, dark blue, occuring quite often with a whitish powder.

Uses: The rootlets of the juniper can be used to make baskets and cordage. The tree can also be used for many of the same purposes described for the cedar. It will even make a serviceable bow if the bow is made wide and thin.

LOCUST
Robina spp.
Bowstaves

MAPLE
Acer spp.
Basket splints

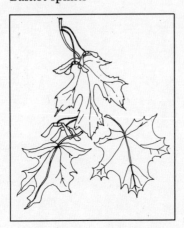

MILKWEED
Asclepias syriaca
Cordage

Description: A stout, hearty plant found along roadsides, fields, and waste places. Grows to 5 feet with opposite, entire leaves. Stem hollow with milky sap. Stem bark green in summer, gray to grayish brown in fall and winter. Leaves medium green above and whitish below, up to 6 inches long. Flower clusters form mostly in the axils of the leaves and vary in color from pale pink to deep rose to purplish. Gray-green seed pods appear from midsummer to fall, shaped like teardrops with elongated stems, up to 4 inches long.

Uses: The milkweed makes an excellent medium-strength cordage that is good especially for Paiute traps and for most bindings. I find it good to use for tying bundles, some small lashing jobs, and for general cordage uses. It is quite silky to the feel and makes a beautiful decorative cordage that will last a long time. It can also be used for the bowstring on a bow drill.

NETTLES
Urtica dioica
Cordage, plaited baskets

Description: A hollow-stemmed weed reaching to 6 feet, found in wastelands, forests, and roadsides across the United States. The stalk and deeply toothed, heart-shaped leaves are covered with a dense coat of stinging hairs. Tiny greenish flowers appear in the axils of the opposite leaves. The male and female flowers are often on separate plants. The stem is 4-sided.

Uses: The nettle makes a fine, strong cordage that is durable enough for use as a light bowstring or for a bow-drill firestarter. It is collected when mature or just after it dies, being careful not to get stung by the hairs. The plants should then be hung in a cool, dry place until dry. The drying also takes away the plant's stinging capacity.

I usually scrape the outer surface of the stalk lightly with my knife or a chip of flint, holding the blade at a 90-degree angle. This removes any of the thin outer covering. The stalk is then broken and stripped. I then like to buff the long fibers lightly over a small branch or across the dull side of a knife. This process separates the fibers and softens them so they are easier to work and bind together.

OAK
Quercus spp.
Bowstaves, baskets, cordage

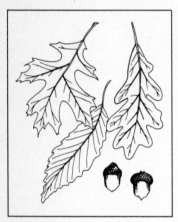

Description: Various species of oaks are found in dry or moist woodlands all across the United States, except in the extreme Southeast. All species make good baskets and marginal cordage. For the best splints, I would recommend the white oak, *Quercus alba.* This is a tall tree, varying in height from 60 to 90 feet. Its leaves are rather evenly lobed, lighter beneath, alternate and simple, and from 3 to 9 inches long. Acorns appear at the end of summer, with cup-shaped capes covering from ¼ to ⅓ of the nut.

Uses: Basket splints can be made from the trunk by soaking for several days and pounding the log with a mallet until the points separate at the growth rings. Oak is a bit more difficult to work than ash for splint-making and should be soaked longer to make it more supple. At critical bends I find that steaming the splints helps to shape the curve of the basket.

Cordage can be made from the inner bark, which can be used for mats and blankets. However, it makes a weak binding. The saplings and twigs can be made into sapling baskets or split in half to make serviceable splints.

The best longbows are made from the wood of oak saplings. Oak is somewhat difficult to read and work and has a lot of "string follow." It also snaps easily if it's not read correctly. But once completed and tested, an oak bow will last a lifetime and hardly needs any fine tuning or recurving.

OSAGE ORANGE
Maclura pomifera
Bowstaves

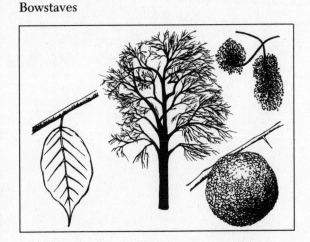

Description: Medium-sized tree, 45 to 55 feet high, found throughout most of the United States due to its widespread use in making fencerows before the advent of barbed wire. Leaves alternate, ovate or oblong, 2 to 7 inches long, untoothed. Flowers form May to June. Fruit is the size of a grapefruit or larger, green with convolutions, appearing in fall.

Uses: Osage orange wood makes one of the finest bows, but it is quite difficult to follow the grain. I would suggest that a beginner not use this wood for a first bow, but save it until later when it's easier to read the grain. Usually the grain has a slight spiral which makes it difficult to use as well as to carve. The wood is yellow and can be used as a dye. At no time should an Osage orange bow be carved or cut, as this will weaken the grain.

PALM LEAVES
Palmae spp.
Baskets, cordage

PINES
Pinus spp.
Cordage, baskets from
needles and rootlets

Description: Evergreen trees found throughout the United States. Leaves appear in clusters of 2 to 5 needles, each cluster bound by a small papery sheath where it attaches to the twig. Branches develop in whorls around the main trunk. Small, pollen-producing male cones are found at the end of the twigs and produce a yellowish dusty pollen in midsummer. Trees bear mature woody female cones.

Uses: Pines can be used for making a variety of cordages and baskets. The soaked roots can be braided into fairly strong cordage, but it's not long-lasting. This kind of cordage is especially useful in a survival situation, but I do not like taking live trees for any reason. Long-dead roots just do not make a good cordage. Baskets can also be made from the small roots and rootlets and will last quite a long time. The needles can be bound together and made into fine pine needle baskets that are quite durable. They can also be made into coil baskets quite easily.

PLUM
Prunus americana
Arrows

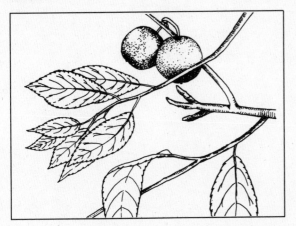

Description: A shaggy bush found throughout the United States. Leaves are 1 to 3 inches long, alternate, toothless, hairless, long-pointed, usually narrow and sometimes wedge-shaped. Flowers in late spring. Fruits reddish or yellow, usually in early fall.

Uses: Plum saplings from any of the *Prunus* species make very serviceable arrow shafts, though some are heavier and more brittle than others. In fact, I find that the wood of most fruit trees is easy to bend and smooth into arrows. Look for straight, smooth saplings that are growing in tough competition with each other.

REED
Phragmites communis
Arrows, mats, baskets

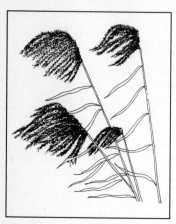

Description: Found in wet areas, either in fresh or brackish marshes. Long, jointed stems, appearing much like bamboo but thin like a pencil. Plumelike flowerheads appearing purplish in summer but turning brown or beige in late summer or early fall. Leaves long, to 2 feet, 1 to 2 inches wide at the base. Usually grows in huge colonies along marsh margins and sometimes out into deeper water.

Uses: The reed makes an ideal arrow shaft, despite the fact that it is very light and shatters easily. It is the fastest arrow known to man, easily outspeeding the finest manufactured shafts. I have shot a reed arrow from a heavy compound bow and a 70-pound pull Seneca bow without the arrow splitting or shattering. The lightning-fast arrow can even penetrate through both sides of a deer and still be found intact.

Unlike other arrow woods, reeds should be gathered when mature and green, though in a survival situation I have used stems almost nine months dead. The shaft can be straightened in the same way as a wood shaft and can be easily prepared and smoothed. In a survival situation, the shaft can also be used without any nock plug or foreshaft. The reed shaft is one of my favorite survival arrows, being especially good for quick, small game or waterfowl.

Leaves of the reed can be collected and woven into serviceable mats for flooring or walls. Leaves can also be used in making baskets of either the coil or the weave type. The stems can be bundled and tied to form thick sleeping mats or to line the entire walls (inner or outer) of the shelter, or for a good source of thatching material. Braided leaves can be used to make thicker mats, much as a coil rag rug is made. All products from the reed are quite strong and durable, making a very comfortable addition to any shelter.

SAGEBRUSH
Artemisia spp.
Inner bark cordage, baskets, mats

SWEET GRASS
Hierochloë odorata
Cordage, baskets

TAMARACK
Larix laricina
Cordage and baskets from rootlets

VELVETLEAF
Abutilon theophrasti
Cordage

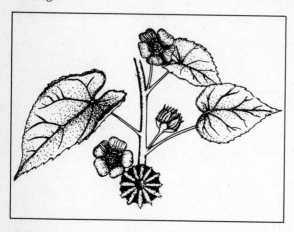

VIRGINIA CREEPER
Parthenocissus quinquefolia
Baskets

WILD CURRANT
Ribes spp.
Arrows

Description: These little shrubs look much like small maple saplings, having 3 to 5 lobed leaves. But unlike the maple, which has opposite, simple leaves, the leaves of the wild currant are alternate and simple. Bark is usually papery and peeling. Many of the *Ribes* species have thorns and are known as gooseberries. Those without thorns are called currants.

Uses: I find that all of the wild currants make good arrows, though some are better than others. The shafts are easy to work and quite serviceable as quick survival arrows. I still have some arrows from the golden currant (a Western species) that were made by Stalking Wolf. These are just as strong and straight as the day he made them over twenty years ago. In fact, my first arrow was made from the American black currant, which I found on my travels with my parents. I strongly suggest that your first few arrow shafts be made from one of the currant or gooseberry bushes.

WILD GRAPES
Vitis spp.
Baskets, cordage

Description: A long twining vine found in thickets and at the edges of wooded areas. Occurs throughout the Northeast and in many other parts of the country, either in wild or cultivated form. Vines are dark and thornless with peeling bark. Leaves are large, often 6 inches across, deeply toothed, alternate and simple—either heart-shaped or lobed at the base. Flowers appear in midsummer and are greenish by late summer. By early fall a dark purple, fleshy fruit appears. Each fruit contains 1 to 4 teardrop-shaped seeds.

Uses: Wild grape vines can be braided into strong cordage or woven into very strong and durable baskets. Once again, I prefer to use the vine on a newly killed grape plant so as not to waste a valuable food source.

WILLOWS
Salix spp.
Cordage, baskets, arrows

Description: Widely varied trees and shrubs, hearty and deciduous, usually found along rivers and wetlands. Leaves are alternate and simple, usually lighter below. Twigs are usually bright green and can be recognized at a considerable distance. Flowers and catkins covered with fine, silky hairs.

Uses: Smaller twigs and branches can be used as sapling baskets. The saplings and suckers can be made into serviceable arrows. The logs, though on the brittle side, can be pounded into basket splints after a lengthy soaking.

The inner bark of some willows makes a fine cordage, though it's not very strong or soft. I have softened willow cordage by soaking it in a lightly salted solution or in brackish water for several hours, bleaching in the sun, and buffing over a small branch until it's lightly frayed.

YEW
Taxus canadensis
Bowstaves, arrows, handles

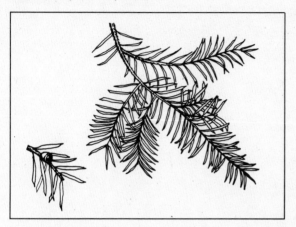

Description: An evergreen shrub found in damp woodlands in Northeastern United States up to Canada. Height to 6 feet. Needles ½ to 1 inch long, green on top and bottom, flattened. Female plants bear a red fruit about the size of a pea. Other useful species include *T. cuspidata* and *T. bacata*.

Uses: The yew tree has long been used as a bow wood in England and other European countries. It is a good bow wood, comparable to Osage orange but much easier to work. It is quite difficult to find a stave long enough or free of knots, but it is quite easy to follow the grain with the tool to work out knots and flaws. Any of the yew trees can be used for bows, some yielding better results than others.

YUCCA
Yucca filamentosa
Cordage, baskets, mats

Description: A hearty flowering plant of the desert and dry places, now found throughout the United States as an ornamental. Has many off-white, bell-shaped flowers that are borne on a woody central stalk. Leaves are thick and lance-shaped, sharply pointed on the ends, arranged in a rosette at the base. The stalk may reach over 6 feet and the leaves over 3 feet. Eastern species have red fibers in the leaf margins. Flowers in late summer.

Uses: Yucca leaves make some of the finest and strongest cordage. Leaves should be collected green or almost dead and dried until supple. They are then buffed over a stick until the fibers separate. Sometimes the leaves must be pounded with a rock in order to separate the cordage fibers. Old leaves can sometimes be used, as long as they are still supple.

The cordage of the yucca leaves is strong enough to be used with the bow drill and even on some of the heaviest bows. The roots and leaves can also be used as basket material or for the making of durable mats, either by braiding or weaving. They are just as strong as the cordage. If woven green, the leaves and roots need not be soaked or softened.

INDEX

As you know from reading this book, sharing the wilderness with Tom Brown, Jr., is a unique experience. His books and his world-famous survival school have brought a new vision to thousands. If you would like to go further and discover more, please write for more information to:

The Tracker

Tom Brown, Tracker, Inc.
P.O. Box 173
Asbury, N.J. 08802-0173
(908) 479-4681
Tracking, Nature, Wilderness Survival School